# HOW STRATEGIC AIRPOWER HAS CHANGED THE WORLD ORDER

*This book is dedicated to the furtherance of an enduring United Kingdom–United States strategic maritime alliance in an uncertain world.*

# HOW STRATEGIC AIRPOWER HAS CHANGED THE WORLD ORDER

## FROM THE 100TH BOMB GROUP IN 1943 TO THE FALKLANDS & BEYOND

COMMANDER 'SHARKEY' WARD & DR ANTHONY WELLS

Pen & Sword
MILITARY
AN IMPRINT OF PEN & SWORD BOOKS LTD.
YORKSHIRE – PHILADELPHIA

First published in Great Britain in 2024 by
Pen & Sword Military
An imprint of
Pen & Sword Books Ltd
Yorkshire - Philadelphia



ISBN 978 1 03610 656 0



A CIP catalogue record for this book is available from the British Library.



Typeset by SJmagic DESIGN SERVICES, India.

Printed and bound in the UK by CPI Group (UK) Ltd.

Pen & Sword Books Limited incorporates the imprints of Archaeology, Atlas, Aviation, Battleground, Digital, Discovery, Family History, Fiction, History, Local, Local History, Maritime, Military, Military Classics, Politics, Select, Transport, True Crime, After the Battle, Air World, Claymore Press, Frontline Publishing, Leo Cooper, Remember When, Seaforth Publishing, The Praetorian Press, Wharncliffe Books, Wharncliffe Local History, Wharncliffe Transport, Wharncliffe True Crime and White Owl.

For a complete list of Pen & Sword titles please contact

PEN & SWORD BOOKS LIMITED
George House, Units 12 & 13, Beevor Street, Off Pontefract Road,
Barnsley, S71 1HN, UK
E-mail: enquiries@pen-and-sword.co.uk
Website: www.pen-and-sword.co.uk

or

PEN AND SWORD BOOKS
1950 Lawrence Rd, Havertown, PA 19083, USA
E-mail: uspen-and-sword@casematepublishers.com
Website: www.penandswordbooks.com

MIX
Paper | Supporting responsible forestry
FSC® C013604
FSC www.fsc.org

# Contents

**Part III: Beyond the Falklands War**

# Foreword

We live today in a turbulent world, presenting difficult and dangerous strategic challenges of several sorts, in which the price of 'getting it wrong' could well be globally catastrophic, even existential. To put it bluntly, although some of these potentially existential challenges are not simply military, there can be no greater moral obligation on our national leaders than the avoidance of a nuclear war that would render all other interests superfluous.

It follows that there has never been a time in which deep, courageous, visionary, and perhaps even iconoclastic, thought about our national grand strategies was more necessary, or in many nations more neglected. By grand strategy I mean that balance of diplomatic, economic, ecological, technical, industrial, social and military power (all of which are of importance and difficulty) which make up the framework within which the various instruments of policy may be deployed. These in turn make up the response of a nation or alliance to the dangerous world we confront. All contributions to this activity are contributions to one of the most crucial and urgent debates in the history of humankind.

In this book, two highly experienced and knowledgeable

Vice Admiral (Retd) Sir Jeremy Blackham KCB.

authors set out to discuss a vital part of the military aspect of the complex tapestry of human co-existence. They present strategic airpower as the vital strategic capability that it is, illustrating it with hard-earned personal experience. They amply demonstrate the significance of this important capability. Personal experiences, appropriately used, can certainly provide lessons but the past must be seen, not simply in the context of the past, but in the light of the insights it offers into the current and future political and military challenges we face. Not everyone will agree with the balance of strategic thought and personal experience struck in this book, but all readers should surely see it as strong advocacy for a capability that, as the book shows, is a vital strategic instrument. The wider and better informed the debate, the greater must be the prospect of finding the right answers we so much need. The authors provide a most useful contribution to that important debate.

# Pre-Publishing Reviews

**By Jeremy Black, author of *Air Power: A Global History*. One of the world's finest scholars of military history.**

'The key role of air power across an important range of spheres arises clearly from this cogent, ably written and highly significant volume. The examples are very well selected, and this book richly deserves attention.'

**By Commander Laon Hulme OBE, Royal Navy Mentioned in Dispatches, HMS *Brilliant*, Falklands War. Two sea commands.**

Lieutenant Commander Hulme receiving Mentioned in Despatches Award from Captain Bob McQueen.

The following extract from a letter written in the heat of the action by Laon to his wife provides a superb insight into the gallantry demonstrated by the frigates and destroyers deployed in the battle for Falkland Sound, 1982:

> I had ordered that we have the aircraft control frequency readily available in the Ops room, more out of interest to know what was going on than anything else and ended up doing two days of fighter control (Fri & Sat) when Antrim had been damaged. A/C under my control splashed 4 Mirages and damaged another 3 Sky Hawks and 1 Pucara. None of the Aircraft under my control were damaged by Enemy aircraft but the no.2 to 'Sharkey Ward', CO of 801 NAS Sea Harriers from HMS *Invincible*, was hit by

HMS *Brilliant* on 21 May. While Sharkey Ward and Steve Thomas were engaging and shooting down three Mirage V Dagger fighter-bombers inbound to the Sound, a fourth undetected Dagger inflicted cannon fire damage to the ship's side and the operations room, narrowly missing Laon but injuring those close to him. He carried on controlling our Sea Harrier aircraft gallantly and without flinching.

ground fire. [He, Steve Thomas, recovered safely to his mother ship.]

In short, in that one day we reduced by 25% the total number of serviceable Arg aircraft! Quite a thought. However, it was not that easy. We succeeded in our aim of defending the very vulnerable landing force but mainly by offering ourselves as the target!

## Commander Hulme's Review

The Falklands War provided a clear demonstration of the UK's political will to recover the Islands after the Foreign Office's failure to appreciate the threat despite the clear drumbeats from Argentina. However, the UK was fortunate as the Argentinean Junta's incursion failed one of the key principles of war – timing – as the infamous Nott Review was actively considering selling the carrier *Invincible*, an integral part of the Royal Navy, to the Australians.

The Royal Navy, which was then still very able to 'project power from the sea' was manned by highly trained, well-motivated, personnel who were able to 'think outside the box' in any combat situation. A task force was duly mounted and, as they say, the rest is history.

Today the RN is less than half the size it was at the time of the Falklands War in 1982 with fewer than seventy-five actively commissioned vessels. This may explain why King Charles' coronation was the first in more than a century not to be accompanied by a Naval Review to showcase both the might and standing of the Royal Navy.

Against the background of the fortieth anniversary of the Falklands War, which was but a 'skirmish' when viewed against the world turbulence being injected by the dictatorships of Russia and China today, Tony Wells and Sharkey Ward's magnum opus reads as a *cri de cœur*, but one is bound to wonder if their well-informed recommendations are just too late.

It is apposite to note that within these covers the authors have shown that US and NATO failed in both their deterrence of Russia invading Crimea and then Ukraine, and the Chinese seizing the key islands of the

South China Sea – despite the presence of the powerful 7th Fleet. These incursions led directly to Russia and China's military expansionism and are clearly both part of their national long-term strategic aims. Their deceptive stratagems, which may well be coordinated, are aligned with Sun Tzu's military philosophy to reduce the West's economic resources while weakening its resistance to fighting so that when war comes it will be swift and decisive and at a time of their choosing.

The authors are calling for visionary change as the economically weakened US shows signs of not wanting to continue to act as the 'world's policeman'. This is certainly timely as Russia's invasion of Ukraine has shown clearly that while NATO is an alliance (and the EU would wish an army of its own) these allied countries, both before the conflict and during, have been acting primarily to benefit or protect their own national interests – often to the detriment of the other allied members.

Can NATO's deterrence posture therefore be based on 'peace in our time' without the nations' willingness to pay for the necessary resources to ensure a timely coordinated response to autocratic aggression – particularly when the aggressor is employing overwhelming force?

Tobias Ellwood, former Chairman of the Defence Select Committee, has opined that both China and Russia are 'openly pioneering a more authoritarian approach to global relations, corrupting the boundaries of acceptable state behaviour'. He has stated that: 'We've clearly entered a new era of insecurity yet remain committed to a peacetime defence budget and that our entire defence posture is woefully inadequate to protect us from this new Cold War.' Visionary indeed.

There is therefore a pressing need to urgently revise how UK defence is conducted and afforded. We can no longer fall back on 'hearts of oak' – our nation and our service personnel deserve better.

**By Lieutenant Colonel Ryan Bachman, United States Air Force, U-2 Squadron Commander, Beale Air Force Base, California**

'Sharkey Ward and Tony Wells bring to life the fascinating insights and daily mortal challenges experienced by American and British combat aviators in two distinctly different wars. Through their stories about two of history's markedly different combat aviation experiences, they

*Right*: Colonel Ryan Bachman.

*Below*: Ryan's redoubtable USAF U-2 'spy plane' completing a typical mission in the south-east Pacific.

demonstrate a common military defence strategy that remains critical to today's and tomorrow's conflicts.

'An exciting read that not only takes the reader into the cockpits of B-17 aircrews flying bombing missions over Nazi Germany, to British Sea Harrier pilots flying attack and combat air patrol missions during the Falklands War, but importantly fuses a national strategy and doctrine (strategic maritime policy) that decision-makers and combat leaders today need to prioritise.

'Before reading this book, I was not aware of the intense opposition that some senior figures in the British Government and the Ministry of Defence had to the UK's Strategic Naval Aviation component.

'As a student of military strategy and a front-line USAF operator myself, I know that the criticality of rapid global reach and presence makes maritime-based combat aviation an absolute necessity for effective deterrence: demonstrating visible power and dominance and having a proud history of combat success. Our land-based air forces and our navies alone can't achieve 100 per cent of this: rather the investment in the layered effects from both starts to define an effective strategy. It's striking to me that the full value of this joint strategic capability could be called into question in Whitehall as much as it seemed that it was (and apparently still is).

'From the aviation enthusiast to students of military history and strategy, to state-level leaders, this is a great book to learn from, to have in the library, and to which to refer.'

**By Christopher 'Monkey' Inman, Major USAF (Ret.), B-2 Mission Commander**

'This book ought to be required reading at the Royal Air Force College and Britannia Royal Naval College. As our shrinking world has eroded the effective distance between Britain and her detractors, so has arisen the military imperative to correct this creeping vulnerability. The authors herein handily persuade that such a perilous strategic imbalance can only be put right through the existence of a full-time, global-reach air and maritime capability. Dr Anthony Wells and Commander 'Sharkey' Ward have coupled their unassailable research with gripping, first-person combat narrative. The result is an informed and impassioned appeal for a robust and durable strategic air capability for the island nation. Whitehall, take notice!'

Chris 'Monkey' Inman demonstrating the B-2 to general officers during their 'Capstone' flag officer training. Andrews AFB, Virginia, *c.*1998. This is aircraft 88-0328, the 'Spirit of Texas'.

Major Inman also flew the B-52 Stratofortress, seen here conducting air-to-air refuelling.

**By Dr Richard Twogood, Formerly Deputy Associate Director for Electronics Engineering at Lawrence Livermore National Laboratory, California**

'This is a critical book that exemplifies the fine working relationship of the Royal Navy and the US Navy in an increasingly challenging world.

'As someone formerly responsible for highly classified programmes, including joint US-UK research and development, I thoroughly recommend this must-read book of great distinction.'

**By Jay Rosenthal, certified consulting meteorologist and former senior civilian US Navy meteorologist and oceanographer, retired**

'This book by Dr Wells and Sharkey Ward explores the necessity of strategically using both air power and sea power together in an integrated fashion to identify, discriminate and attack targets in global conflicts from the bottom of the sea to the top of the atmosphere. With their incredible knowledge and bravery from World War II up through the Falklands conflict and beyond, they knit C4I, Intel, expertise in radar and surveillance and simple logic to identify how strategic thinking and integration of the natural and operational environment can provide lethal success in battle.'

# Preface

The two authors of this book are acknowledged as having possibly unmatched, in-depth expertise and hands-on experience in their specialist maritime warfare areas of interest: Dr Anthony Wells in naval intelligence at the highest level on both sides of the Atlantic, and Commander 'Sharkey' Ward in naval air warfare. Short biographies are provided at the end.

The book reaches across time and briefly describes eight decades of significant strategic airpower-related events, whether land-based or naval, that:

- underwrite the continuing benefit of the special military relationship between the United States and the United Kingdom, and that,
- in many cases, reveal a lack of investment wisdom in the UK and a less than fruitful allocation of operational control of and/or airpower deployment in the field.

It is an untold story of the uneasy progress of naval air warfare development within the UK and its synthesis with land-based airpower.

These two strands of airpower are mutually supportive in the context of US and UK global security interests. But the book shows clearly that the full benefit of such mutual support has often been hampered by the disregard of those in power concerning the strategically important air warfare and maritime warfare lessons learned during the last eighty years. That disregard has resulted in a steady decline in the strength and global capability of the Royal Navy, Royal Marines and the Royal Fleet

Auxiliary. This does not fit well with the astute warning delivered by elder statesman Winston Churchill:

> Whereas any European power has to support a vast army first of all, we in this fortunate, happy island, relieved by our insular position of a double burden, may turn our undivided efforts and attention to the Fleet. Why should we sacrifice a game in which we are sure to win to play a game in which we are bound to lose?

The winning game he referred to was the proven and traditional island nation commitment to strategic naval power, through which Britain had long secured its homeland and global national interests. He could see that land-based air power in any form could not provide visible deterrence and security throughout the global commons, e.g., it was the UK's formidable naval strength and visible presence (not the Battle of Britain) that deterred Hitler from attempting to invade Britain across the Channel with his ill-conceived Operation Sea Lion.

Churchill's warning was not heeded on two key occasions in the UK's recent military history.

The first was the fateful Dennis Healey decision in 1966–67 to remove aircraft carriers from the Royal Navy fleet and cancel the build of the new strike carrier CVA-01. This was at the behest of the RAF, which misled the minister, ignoring all recent naval history and persuading him through misinformation that they could carry out the strategic roles of aircraft carriers throughout the global commons. The result of this decision was the inevitable erosion of the UK's strategic military and political power projection capability and it signalled to the Argentinians an opportunity for them to invade the Falkland Islands in 1982, the last UK strike carrier, HMS *Ark Royal*, being withdrawn from service in 1977.

The second, which contributed to the same result, was the ludicrous statement by UK Defence Minister John Nott prior to the Falklands conflict, who posed this question to his successor, Michael Heseltine:

> … are we to have a Navy which is appropriately designed to fight low-level wars against the Hottentots and a Navy which is to be appropriate if need be for invading the uninhabited

The multi-role air defence fighter, the Phantom F-4K, launching from the waist catapult of HMS *Ark Royal* during its final commission.

> islands like Anguilla, or are we to have a Navy which is actually designed to be [meet] the main threat of the British people … Which is principally 35–40 miles off the coast of Britain and the nation's flank from Soviet threat.
>
> Taken from Churchill College Archives

This ignorance, displayed by Nott, of the then major, long-range, Cold War threat from the Soviets and of the global utility of robust naval air power, led to the announcement by the UK government just prior to the Falklands War that it was to sell HMS *Invincible*, UK's new Sea Harrier carrier to Australia. This was seen as a 'welcome mat' by the Argentina dictator, Leopoldo Galtieri, who proceeded with his invasion of the islands.

That invasion is a story that has been well told in detail by many different authors. This book intentionally briefly summarises such key events and presents the reader with a broad historical understanding of why the UK government has now publicly returned to a strategic maritime policy for its armed forces; as indicated with the statement by the then Secretary of State for Defence, Philip Hammond, that 'our new Strike Carriers are at the heart of our Strategic Defence Policy'.

The actual events of recent military history cannot be changed but have often been misinterpreted to suit the opinion of third-party researchers and observers, rather than taking the events at face value and recommending lessons to be learned from them. It is the book's intent to follow the latter course, and in doing so, it suggests firmly that the UK government should channel a higher proportion of the available defence budget into global maritime utility, mobility and presence. Such a change would save the UK taxpayer money while increasing the fighting efficiency of the UK's armed forces, particularly the Royal Navy and its embarked airpower.

With tensions rising significantly in the Indo-Pacific region, it emphasises that allied strategic airpower is deployed regularly to Britain, especially during times of international tension, courtesy of the USAF B-1, B-2 and B-52 aircraft. The UK's new Carrier Battle

A B-52H Stratofortress arrives in Guam, July 2023. (Photo by Airman First Class Nia Jacobs)

The B-1B supersonic strategic bomber.

Groups also represent strategic maritime airpower. This book suggests that investment in large fleets of short-range, land-based tactical fighter aircraft is misguided, unjustified and should be discontinued.

This decades-long story is split into three parts.

**Part I:** The Second World War and the 100th Bomb Group. Dr Wells provides the main body of this section. It is an intimate insight into the USAF's participation and its major contribution to Allied victory with the daytime precision strategic bombing of Nazi Germany, targeting military installations.

This was achieved at great cost in blood and treasure, turning the tide against the Nazi war machine. In parallel, the UK's Lancaster bombers conducted night-time carpet-bombing raids against population centres that did little to limit the military resources of the German war effort. Part I also highlights the unfortunate reluctance of the UK's air force to

The 100th Bomb Group B-17s in action. (© David Julian)

allocate much-needed resources to counter the German U-boat offensive, which nearly brought Britain to its knees.

**Part II:** The Falklands Air War. The history of the conflict is intentionally related in the first person by Commander 'Sharkey' Ward, who had been given full authority and control over preparing the Sea Harrier for combat operations.

It provides a detailed summary of the CVA-01 cancellation decision, engineered by RAF dishonesty, and of the subsequent difficulties faced and overcome by the Royal Navy's new air defence fighter, the Sea Harrier. With very limited numbers against great odds (including high command ignorance of its combat capabilities), the little jump jet provided enough protection to the UK Carrier Battle Group and the Amphibious Warfare Group to successfully facilitate the recovery of the Islands by our Royal Marine-led ground forces – a convincing demonstration of maritime strategic power.

Sabre rattling by the Argentinians concerning their unjustified claims to the sovereignty of the British Overseas Territory, the Falklands, is an ongoing affair and now they receive strong support for their case from Russia, Germany and among others, China, and even the European Union. In the light of recent world events, including the Ukraine War and the Taiwan stand-off, the South Atlantic has assumed strategic significance. Chinese leader Xi's blatant desire for territorial aggrandisement and influence has included overtures to the Argentinian government with an eye to gaining a military foothold there and on the islands, and we in the West must thwart this desire through the execution of the UK's declared strategic maritime policy – and in coordination with United States Navy. The unilateral cancellation of the Foradori–Duncan pact by Argentina in early 2023 should be ringing alarm bells in Whitehall. Britain ignored the warning signs prior to Galtieri's invasion in 1982 and we must not be caught with our pants down once more.

**Part III:** Beyond the Falklands War. This section, authored by Dr Wells, provides valuable insight into the major strategic significance of the USAF presence within the UK throughout recent decades. This presence, filling critical roles with extensive capabilities, provides the global military reach that is not otherwise available from UK land-based air due to budget constraints. It suggests that there are many other roles that can be conducted effectively and cost-efficiently by the Royal Air Force, without arguably unjustified major investment in large fleets of land-based, short-range, tactical fighter aircraft. It suggests that the latter have little, if any, strategic utility for global expeditionary military operations and that funding for the same should be diverted to more threat-related investment in support of the UK's 'alliance-based' maritime strategy, the principal component here being the Royal Navy Fleet and its embarked airpower. This would enable a critical and robust power projection capability to be realised in support of our major ally, the USA, and would be an order more cost and operationally effective. That would be beneficial to the taxpayer in this time of extreme fiscal constraint and would be welcomed by all our allies as our world faces increasing international tension and challenges.

# Introduction

The role of air power has gone through many changes since the Second World War. Driving factors have been the international security environment since the end of the Nazi era, advances in technology, the various key alliance structures that have been created in response to perceived threats, and the individual national security needs of the major powers. Air and space warfare join underwater warfare, characterised by the development of nuclear-powered attack and strategic submarines, as most likely the pre-eminent domains in which the lead post-Second World War nations have made the most strides. These have impacted the global security environment and continue to do so.

Air warfare developed into quite separate and very different land and maritime domains. They grew in parallel, with unique characteristics that covered both the tactical and strategic environments. The latter has been the driving force for national and alliance-based technical and operational developments and the huge investments associated with complex acquisitions.

This book concentrates on the special and unique relationship of the United Kingdom and the United States. It addresses how together they focused and combined their national security interests at the high point of the Second World War and then transitioned over successive decades their individual and joint security needs into aviation programmes in the land and maritime applications of air power in the service of their joint and national interests.

It is important at the onset to understand the fundamental differences in the application of air power from land and from the sea. There are self-evident differences. These are quintessential in understanding why and how over the period being considered, from early 1942 (when the US Eighth Air Force arrived in the United Kingdom to initiate daylight strategic air warfare against Nazi Germany) through the United Kingdom's successful Falklands' campaign in 1982, to today's very much changed and demanding international security environment.

Maritime air warfare is at the simplest level conducted from ships carrying aircraft and helicopters and, in addition increasingly today, a range of unmanned aerial systems. The aircraft carrier has been and still is the principal platform for launching and retrieving manned and unmanned aircraft at sea. It provides capabilities that include flexibility in location, ability to move expeditiously from one sea area to another, and can be forward deployed into any area of the world's oceans in accordance with the established laws of the sea.

Aircraft carriers can conduct strikes from the sea against land and sea targets and they can conduct deterrence, blockade, and containment missions by the very nature of their presence, and forward deployment, as was witnessed for example during the Cuban Missile Crisis in 1962. And more recently, when Iran threatened to close the Hormuz Strait to merchant shipping, it rapidly changed its tune when a US Navy Carrier Battle Group entered that arena.

The United States nuclear-powered aircraft carriers (CVNs) are quintessential platforms because they have no refuelling needs and can operate indefinitely and at high speeds over lengthy periods by being resupplied at sea with full logistical support, including weapons resupply. The British aircraft carriers, being non-nuclear powered, require refuelling at sea from support tankers. The deployment flexibility of these aircraft carriers, plus other ship types carrying, for example, US Marine Corps vertical and short take-off and landing (VSTOL) aircraft, and assault helicopters, provides both country's leaderships with a wide range of options at the strategic and tactical levels.

Maritime air power has offered enormous flexibility and strategic adroitness since the United States Navy, Marine Corps, and the British Pacific Fleet demonstrated in the war in the Pacific how a maritime war can win back vast areas of territory and retrieve and re-establish other vital national interests. The latter include the freedom of the seas, the passage of global trade, and support to key allies and trading nations threatened by opposing powers.

Aviation related to land warfare has a different set of characteristics. It has been focused on national security interests, e.g. homeland air defence, which are very different from the global reach and mobility of maritime air warfare.

Land-based aviation requires fixed runways and logistics. These do not move and require a different set of protective assets from naval aviation at sea. Both require protection in very different ways. By the nature of their fixed sites, land-based aviation is controlled by and limited to geography, the combat range of aircraft types, and the ability to provide in-flight refuelling to enhance range and deployment times. Overflight rights may both enhance or limit operations and alliance-based, non-national US and UK facilities are subject to political change. Air space over land has different peacetime requirements from, for example, international airspace over the oceans of the world. Air traffic control authorities play key roles in the ways in which land-based aircraft can fly over sovereign territories. This applies not just to the separation of civil and military aircraft for safety purposes. The political aspects have a separate dimension. This was illustrated in 2022 when Germany restricted Royal Air Force flights flying military supplies to Ukraine in German airspace. This occurred between friendly NATO allied nations. Other situations between non-aligned or non-friendly nations present greater challenges.

Maritime air defence for aircraft carriers and associated platforms, such as accompanying destroyer and frigate escorts, has very different attributes and characteristics from the land warfare air defence of fixed runways and air base facilities. CVN and CV air defence tends to be intrinsically associated with their strike operations. For example, the US EA-18G electronic warfare variant, the Growler, provides defence suppression, both strike support and electronic air defence for a US carrier-based Battle Group or Expeditionary Strike Group. This is not always available with land warfare because of geography and the ways in which sovereign territories are distributed. Some may include allies, some may be neutral or non-committed, and others may be hostile. Land over-flight rights may be significantly inhibiting for air defence operations from land bases.

Intelligence gathering at sea has historically been more flexible and adaptable to technical change than land warfare and its aviation component. Maritime assets can collect intelligence, distribute, and act upon it in more advantageous ways than their land-based counterparts, who are restricted by national boundaries and in situ counterintelligence across non-friendly borders. Submarines have provided outstanding covert intelligence support to both US and UK aircraft carriers. In addition, sea-based electronic warfare platforms have and continue to provide round-the-clock support to maritime air operations. Land-based aircraft can perform similar intelligence operations but have territorial restrictions and may require space-based intelligence support in situations where penetration of national sovereignty is clearly either highly risky or prohibitive.

The above is a general overview. The following chapters will explore and illustrate in detail how the US and the UK have evolved in the land and maritime air warfare environments.

This will be done not by just revisiting and restating well-established history and descriptions of well-known events, aircraft systems, and outcomes. The initial chapters will examine Second World War land warfare through the lens of one Air Group within the United States Eighth Air Force. Later, maritime air warfare will be presented and analysed through the experiences of the Royal Navy and Royal Marines in the 1982 Falklands Campaign. In parallel, commentary will be made throughout on the relative differences between land and maritime air warfare and, in addition, US and UK approaches will be compared and contrasted. Underpinning the latter will be constant reference to the two critical differentiating regimens, strategic and tactical. The goal will be to obtain insight into the best ways that both nations, together and individually, can pursue current and future air warfare policies and programmes that truly meet national security needs.

The third part will build on the prior two parts and address the current and future land and maritime air warfare needs of the US and the UK. This part will address all aspects as best understood in both countries, including the current political landscape, individual national resources, and how the long-standing relationship between the US and the UK can serve to establish thoroughly well-conceived, planned, and executed land and maritime air warfare capabilities that serve the mutual and national interests of both nations.

The Second World War, the Cold War, the war on terror, and events more recently in Europe, the Middle East, and Asia provide the overall background. The detail in each of the three parts is intended to provide an accurate synthesis and overview that will permit readers to be both well informed and draw reliable conclusions about how best the US and the UK can work together to form a coherent policy and plans for future air warfare. The latter is too important to be addressed in anything less than a fully accurate and professional way. The national security of the United Kingdom and the United States deserves no less.

That said, it will become clear to the reader that the West's strategic airpower is vested in:

- the land-based Strategic Bomber Force of the United States Air Force, and equally in
- the fighter and strike aircraft embarked in the aircraft carriers of the United States Navy, the United Kingdom Navy and those of our allies.

# Part I

# The Second World War and the 100th Bomb Group

# Chapter 1

# The 100th Bomb Group Arrives in the UK

On 31 May 1943, 140 pilots, co-pilots, navigators and bombardiers flew to the United Kingdom from the United States to form the initial core of the United States Army Air Force's 100th Bomb Group within the Eighth Air Force. They were based at the Royal Air Force base at Thorpe Abbotts, east of Diss in the English county of Norfolk. Today the original control tower forms the 100th Bomb Group Memorial Museum, which is open to the public.

The 100th Bomb Group comprised the 349th, 350th, 351st, and 418th Bomb squadrons. These squadrons constituted a strategic bombing organisation until the war ended in 1945.

Their core mission differed considerably from that of the Royal Air Force's Bomber Command. The 100th Bomb Group was a truly key 'strategic' organisation employing 'precision daylight' bombing of critical German strategic targets. The impact of the 100th together with the other bomb groups of the US Eighth Air Force dramatically changed the course of the war together with other key capabilities, not least the code-breaking capabilities of Bletchley Park in the UK and Station Hypo in Hawaii. The 100th destroyed critical German infrastructure, war machine manufacturing, distribution, and operational centres. Before the arrival of accompanying US fighter aircraft, the legendary P-51 Mustang, to escort the B-17 bombers of the 100th and other Eighth Bomb groups into Germany and other key targets in Nazi-occupied Europe, the 100th suffered heavy losses. This extraordinarily valiant group acquired the sobriquet the 'Bloody 100th' because of its early heavy losses attacking critical strategic targets. Other Eighth groups suffered overall greater losses. The 100th's losses were renowned because of the heavy losses in specific key strategic operations.

The 100th will be viewed through the lens of one individual navigator who survived and lived a long life. His son has kindly provided his personal records. Paul F. Julian of the 351st Squadron of the 100th Bomb Group will represent the detailed look at the overall major achievements of the US Eighth Air Force in the Second World War.

The Royal Air Force had, by contrast, a different approach to air warfare over Germany and occupied Europe. The legendary Air Chief Marshal Sir Arthur 'Bomber' Harris was the architect and leader of the Royal Air Force's Bomber Command. After the Battle of Britain in the summer of 1940 Bomber Command began major offensive operations against Nazi Germany.

It is important to recall the dire situation that the United Kingdom faced before the Eighth Air Force arrived in 1942. Until the attack on Pearl Harbor on 7 December 1941, the British stood alone. After the United States' Declaration of War against Japan and Germany, President Roosevelt was able to increase direct support to the UK. The US had provided Lend-Lease support, vital oil and food imports to Britain, and various discreet weapons support via Canada before Pearl Harbor. These were critical to help keep the UK alive and well at a time when the German Kriegsmarine's U-boat attacks were creating havoc in the North Atlantic.

The years 1939, 1940 and 1941 were extraordinary valiant years for the British, standing alone, fighting off a possible invasion, withstanding intense bombing of its cities, Royal Air Force airfields, Royal Navy bases, and on its industrial base and war production machine.

Until the arrival of the Eighth Air Force, Bomber Command offered a national palliative to Nazi aggression, typified by Operation Chastise, commonly known as the Dambusters Raid of 16–17 May 1943 by 617 Squadron. The Barnes Wallis-designed 'bouncing bombs' were used in the attacks on the great German dams of the Ruhr valley. The Möhne and Edersee dams were breached, causing serious flooding of the Ruhr valley and villages in the Eder valley. The Sorpe Dam sustained only minor damage. Two hydroelectric power stations were destroyed and several more damaged. Factories and mines were also damaged and destroyed. It is estimated that 1,600 civilians died – 600 Germans and 1,000 forced labourers, mainly Soviet – in the flooding that followed the breach of the dams.

The commander of the Dambusters Raid, Wing Commander Guy Gibson, awarded the Victoria Cross for leading this incredibly courageous attack, became legendary. However, the raid came at considerable cost. Of the nineteen Lancaster bombers that set off for Germany, eight were shot down, with fifty-three aircrew killed and three taken prisoner.

Albert Speer wrote in his book, *Inside the Third Reich*, the following: 'That night employing just a few bombers, the British came close to a success which could have been greater than anything they had achieved hitherto with a commitment of thousands of bombers'. Speer was amazed that the British did not follow up to prevent reconstruction of the dams. Barnes Wallis himself was also of this view. He expressed deep frustration that Harris' Bomber Command never sent a high-level bombing force to hit the Möhne Dam while repairs were under way. He argued that extreme precision, for which the Eighth Air Force and the 100th Bomb Group became renowned, was unnecessary and that a few hits by conventional high-explosive bombs would have prevented the rapid repair of the dam that was quickly undertaken by the Germans.

The strategic underpinnings of the Dambusters Raid and the other major night bombing raids by the Royal Air Force were conceived in order to draw German air resources away from the UK and back into Germany, culminating in the Berlin raids of the winter of 1943–44. In 1944 it meant clearing the way for the air warfare aspects of Operation Overlord, the D-Day invasion of 6 June 1944, and the assault on Germany that followed the successful landings.

Bomber Command's night-time carpet bombing of German population centres following the devastating 'Blitz' by the German Luftwaffe on British cities, airfields, and naval bases provided considerable propaganda and morale benefits for the British people. But the unforgiving and savage assault on civilian targets did not have the desired effect on either country's war machine and, in today's world, would be considered war crimes, as with Putin's missile assault on Ukrainian cities and towns.

However, there were other aspects of RAF Bomber Command's German offensive operations at night that raised controversy, detailed analysis, and discourse within a very small, technically elite, and discreet group early on in the war. Notwithstanding the Dambusters Raid the vast majority of Bomber Command's night-time operations were non-precision, non-strategic, and what were termed 'saturation bombing raids'. The casualty consequences were severe.

RAF bomber command aircrews suffered heavy losses. Of an estimated total of 125,000 RAF aircrew, 57,205 were killed. This constituted a 46 per cent death rate. In addition, a further 8,403 were wounded, and 9,838 became prisoners of war.

Were there alternatives to these heavy, devastating losses of the flower of British young men? The answer to this lies at the strategic level.

Until the Eighth Air Force arrived in 1943, RAF Bomber Command had been caught in a closely guarded and secret controversy over its strategic objectives, and Bomber Harris' insistence on pursuing night-time raids over Germany with their massive losses.

Britain was almost brought to its knees by the devastating attacks by German U-boats on merchant shipping bringing vital war supplies, food, and oil to the besieged country. The years 1940–43 were critical, with the British strategic position ameliorated by the entry of the United States into the war after Pearl Harbor in December 1941. It took until 1942–43 for the UK-US allies to mount effective offensive operations against the U-boats, and into 1944 and 1945 to see the final demise of Admiral Karl Dönitz's U-boat force.

The man who later became Lord Blackett led the intellectual and policy debate of the relative merits of Bomber Command night-time operations over Germany versus putting more RAF resources and personnel into the fight against the U-boats in the crucial years 1940–43 and into 1944.

Patrick Maynard Stuart Blackett OM CH FRS (18 November 1897–13 July 1974) was the father of operations research in the United Kingdom and led major contributions during the Second World War that contributed significantly to the Allied victory. He was trained initially as a naval officer, for two years at the Royal Naval College, Osborne, on the Isle of Wight, before moving on to Britannia Royal Naval College, Dartmouth. From 1914 at the outset of the First World War, as a midshipman, he began to see considerable action, including in 1916 at the Battle of Jutland on board HMS *Barham*. While on board *Barham*, Blackett was the inventor of a gunnery device that the British Admiralty patented. After the First World War ended in 1918 the Admiralty sent Blackett to Magdalene College, Cambridge. He left the Royal Navy and began studying mathematics and physics, graduating in 1921. In 1923 he became a Fellow of King's College, Cambridge, a position he held until 1933. His academic advisor was Ernest Rutherford, and Blackett became immersed in research at the Cavendish Laboratory. Before the Second World War he worked at Birkbeck College, University of London, and from 1937 at Manchester University as the Langworthy professor.

During the Second World War, Lord Blackett was in the key position of Director of Operations Research at the Admiralty, working with Room 39 in British Naval Intelligence. He worked closely with E.J. Williams on the most effective ways to ensure the survival of convoys. As a result of this work Blackett obtained great knowledge and insights using sound scientific data and reasoning for greater resources to be used in the countering of U-boats versus the heavy investment in resources and manpower for the night-time area bomber offensive of the RAF's Bomber Command.

All this brought Blackett and his colleagues into confrontation with Bomber Harris and the advocates for the night-time offensive that cost so many lives.

Blackett approached the issues scientifically and from a strategic vantage point. What were the best national security interests of the UK at the time of the devastating impact of U-boat attacks in the North Atlantic? He presented the case and did not win.

However, after the war, the Allied Strategic Bombing Survey proved Lord Blackett correct.

In 1940 he was awarded the Royal Medal of the Royal Society, having been elected a Fellow of the Royal Society (FRS) in 1933. In 1948 Lord Blackett was awarded the Nobel Prize in Physics and, in 1956, the Copley Medal.

In 1953 he became head of the Physics Department at Imperial College, University of London, and retired in July 1963.

Lord Blackett's contribution to the Battle of the Atlantic was outstanding. He remains to this day a giant in the field of military and specifically naval operations research.

What he had ideally sought was greater RAF assets created and assigned to the anti-U-boat campaign. He was totally frustrated by the air force's response.

This issue was exacerbated by the ways in which British airborne maritime reconnaissance and anti-submarine roles were organised and conducted.

There had been serious disagreements between the Royal Navy and the Royal Air Force in the inter-war period over the roles, missions, investment and acquisition of maritime air power. RAF Coastal Command was created in 1936 at the same time that the RAF was restructured into Fighter and Bomber commands. Coastal Command was the third command created.

Royal Navy and Royal Air Force interests became diametrically opposed and in conflict. The RAF prioritised its bomber and fighter forces as the primary deterrent forces. In 1940 the First Lord of the Admiralty referred to RAF Coastal Command as the 'Cinderella Force'. Its headquarters were moved from their initial location in Lee-on-the-Solent to Northwood in north-west London, from where today RAF maritime air operations are conducted at the Royal Navy's Fleet headquarters.

Coastal Command did outstanding work with the resources it had in the protection of Allied convoys against both U-boat and Luftwaffe aerial attacks. Defending the supply lines in the North Atlantic was critical and it became very clear to Lord Blackett and his colleagues in the Admiralty that the RAF's commitment and resources were inadequate. Resources were being absorbed by RAF Bomber Command in a mission that Lord Blackett considered against the vital national interests of the United Kingdom when its very lifeblood was threatened by German U-boats.

In due course the United States Navy came to the rescue after Pearl Harbor.

The contrast between how the United States Navy and how the British organised and managed their respective approaches to airborne reconnaissance and anti-submarine warfare are in stark contrast, and to a certain extent these differences persist in large measure today. The hugely capable Boeing P-8 Poseidon long-range maritime reconnaissance and anti-submarine aircraft that is the US Navy's land-based aircraft has been acquired by the Royal Air Force and operated from northern Scotland. To the US Navy it seems extraordinarily at variance with Royal Navy and UK national interests for its one critical maritime air reconnaissance and anti-submarine capability to be under the command and control of a separate service. The answer clearly lies in the history of Coastal Command and also politics. The Royal Air Force has clearly wanted to retain control.

In 1943 Coastal Command received from the United States the Consolidated B-24 Liberator. This aircraft changed the nature of Coastal

Command operations and successes against the U-boats. Coastal Command saw action from the very first months of hostilities until the end of the Second World War, completing approximately 1 million flying hours, 240,000 operations, and destroying 212 U-boats. Coastal Command lost 2,060 aircraft to all causes. During 1940–45 the command sank 366 German transport ships and damaged 134. It rescued 10,663 people, comprising 5,721 Allied crew, 277 enemy personnel, and 4,665 non-aircrew, and lost 5,866 personnel killed in action. This number is in stark contrast with the huge number killed in Bomber Command.

Coastal Command's functions continued during the Cold War. In 1969, it became part of a newly formed RAF Strike Command that also absorbed the former Bomber, Fighter, and Signals commands. Later, Air Support Command, the former RAF Transport Command, was absorbed into Strike Command.

The issues and politics associated with the control, leadership, acquisition, and funding of British long-range maritime reconnaissance and anti-submarine warfare aircraft and operations persist to this day. The history is clear cut. The current and future will be addressed in later chapters.

The Eighth Air Force arrived in the UK in 1942 after the UK had stood alone and while the above issues were still very much not in the public domain and even within the inner sanctum of British wartime defence policy kept watertight.

The leadership of the Eighth Air Force in Britain during the war is a critical factor. The Eighth was activated at Savannah, Georgia, on 28 January 1942, coming to life at Langley Field, Virginia, on 1 February 1942. Headquarters in the UK were later established in the Wycombe Abbey School for Girls.

The Eighth had outstanding leaders in men such as Carl A. Spaatz, Ira C. Eaker, Jimmy Doolittle, and Curtis LeMay. The 'Mighty Eighth', as it became known would, by mid-1944, have a total strength of 200,000 people, and would be sending more than 2,000 four-engine bombers and 1,000 fighters on a single mission against enemy targets in Europe. The Eighth's highly successful daylight strategic bombing campaign against Nazi Germany came at a price. The Eight suffered about half the casualties of all US Army Air Forces (the United States Air Force, USAF, was formed in 1947), 47,483 out of 115,332. The Eighth won 17 Medals of Honor, 220 Distinguished Service Crosses, and 442,000 Air Medals. The Eighth's combat record shows 566 aces (261 fighter pilots with 31 having 15 or more victories and 305 enlisted gunners), over 440,000 bomber sorties to drop 697,000 tons of bombs, and over 5,100 aircraft losses and 11,200 aerial victories.

The continuity of the roles and missions of what became the USAF after the war in both the UK and Europe will be an important theme in later chapters. Today the USAF in the UK continues to play a pivotal role in the overall strategic picture. It has significant UK bases and enormous capabilities today and for the foreseeable future.

The above has to be seen in the context of British defence policy and the overall global strategic situation. The roles and missions of British naval aviation have to be set not just against the roles and missions of the Royal Air Force. Both have to be set against the extraordinary technical and order of battle capabilities of the USAF in land-based air warfare roles and missions. This is a critical factor as the USAF today has both a presence and capabilities subsuming that of the RAF when the UK is seeking to extend its global reach by a maritime strategy that is underpinned by its two aircraft carriers.

The Eighth Air Force established both this tradition and capabilities in the Second World War. The 100th Bomb Group is the exemplar of that tradition and those capabilities.

In the context of British domestic politics and economics, the defence budget has been sustained by a policy commitment to an alliance-based approach. The UK now and in the future will have to make serious choices about resource allocations in the context of its alliance-based global commitments and domestic industrial base needs.

A maritime strategy requires a major commitment to naval air warfare and capabilities. The United States Air Force can continue to provide a long-term major commitment to a land-based air warfare presence in the United Kingdom. The United States' major commitment to supporting Ukraine against Russia is indicative of the American resolve to support a global strategy to sustain peace and security for those allied and other nations that sustain democratic values and institutions.

A continuing powerful land-based USAF presence permits the United Kingdom to implement a global maritime strategy.

# Chapter 2

# The 100th Bomb Group Begins Intensive Strategic Bombing

On 31 May 1943, 140 pilots, co-pilots, navigators, and bombardiers of the 100th Bomb Group flew to England. This was a historic moment for several reasons, not least the beginning of a commitment and relationship that endures to this day. The United States Eighth Air Force within the US Army Air Force, the predecessor to the United States Air Force established in 1947 at the same time as the Central Intelligence Agency (CIA), was the precursor to a permanent presence in the United Kingdom of major United States airborne capabilities. During the 1990s there were approximately 100 US bases in the UK. Today there are thirteen that cover a wide range of air and air intelligence-related operations. They use Royal Air Force (RAF) designations and include: RAF Lakenheath, RAF Croughton, RAF Digby, RAF Welford, RAF Fairford, RAF Feltwell, RAF Upwood, RAF Barford St John, RAF Fylingdales, and RAF Menwith Hill.

While most of these bases are named Royal Air Force stations, they are leased by the United States for the sole purpose of USAF operations. Although the physical buildings are the property of the UK Ministry of Defence very little that transpires at these bases is controlled by the British government. However, there persist today the same intrinsic relationships that were established by the United States Eighth Air Force in the war, and one of its key bomb groups, the 100th. Critical intelligence gathering and surveillance operations are conducted from some of these bases, and such activities operate within the remit of the historic US-UK intelligence relationship and the wider Five Eyes intelligence community.

Today RAF Lakenheath is the largest USAF base in the UK, hosting the 48th Fighter Wing. This is a global response unit. Lakenheath operations have been launched into Libya in 1986, as well as Iraq and Afghanistan.

RAF Menwith Hill in North Yorkshire, a US National Security Agency (NSA) facility, is a large intelligence centre. The powerful radar system at RAF Fylingdales, also in North Yorkshire, supports the US-UK Ballistic Missile Early Warning System (BMEWS). Space activities and missile interception are also managed from Fylingdales.

In addition to operating airfields for humanitarian and military missions and the above intelligence and surveillance operations, several of these bases also perform training, refuelling, and munitions storage functions. The bases operate within UK-US agreements and laws. The traditions and operations of the Eighth Air Force and the 100th Bomb Group endure.

In 1943 the 100th began operations against several of the most strategic targets in Nazi Germany and other parts of occupied Europe. Before the destruction of German anti-aircraft capabilities, including the Luftwaffe's fighter force, and before the arrival of the North American P-51 Mustang long-range fighter that could escort Eighth Air Force bombers to Berlin and other targets and back, the 100th's squadrons were highly vulnerable and casualties were high. There were two distinct phases of the United States' strategic bombing campaign, pre-P-51, and post P-51. What was termed the 'Little Friend with a Big Impact' enabled the Eighth Air Force to establish complete control of the air, and eventually drive the Luftwaffe from the skies of Europe.

The P-51 prototype was ready on 9 September 1940, first flew on 26 October 1940, and became one of the great accomplishments by the North American company in the war.

In the autumn of 1942, the Americans and British experimented with the Mustang by adding the Rolls-Royce Merlin engine to the airframe, replacing the American-built Allison. The latter was effective at lower altitudes though suffered from a drastic drop-off in performance higher up. With the Merlin engine, the Mustang's top speed was increased to over 400 miles per hour. The first P-51 units arrived in Europe towards the end of 1943, a blessing for the 100th Bomb Group that was suffering frightful losses in its daylight raids. The Lockheed P-38 Lightning that was also a bomber escort was inefficient and not economical. By contrast the Mustang with its high speed and long range and, most of all, its six .50 calibre M2 Browning machine guns, made it the ultimate fighter to escort the 100th and the other groups into Europe.

North American P-51D-5-NA Mustang #44-13926 from the 375th Fighter Squadron. This aircraft crashed on 9 August 1944, and the pilot was killed.

The results speak for themselves. On 6 March 1944, when the campaign for air supremacy kicked into high gear, the Eighth Air Force made its first daytime bombing raids on Berlin. More than 800 United States bombers escorted by more than 900 US fighters attacked Berlin. The Luftwaffe's fighters were drawn into the skies, where the P-51 destroyed them and in due course established air supremacy. By the end of 1944, fourteen of the fifteen fighter squadrons of the US Eighth Air Force were composed of P-51 Mustangs.

The P-51 dominated air combat over Europe, destroying nearly 5,000 enemy aircraft, a prodigious total. By the close of its production lines a total of 15,000 Mustangs were manufactured.

By any standards the P-51 remains the iconic fighter of the war alongside the British Spitfire and Hurricane.

However, until the P-51s joined the 100th Bomb Group in its strategic bombing missions there was a story of indomitable courage and resilience underpinned by superb leadership at the 100th Group and squadron levels. Outstanding leadership in the face of often overwhelming odds was required and produced by the 100th. Its sobriquet, 'The Bloody 100th' was earned because of the high casualties in the pre-P-51 period when the 100th attacked and destroyed several of the key German strategic targets.

On 8 and 9 October 1943, the losses over Bremen and Marienburg were substantial. Four months after the arrival of the 100th's crews 'only three were left on flying status'. Casualties were that high.

However, the damage inflicted by the 100th was hugely significant. The shipyards of Vulkan, AG Weser and Atlas Werke, together with the Valentin submarine pens, the oil refineries, and the aircraft works of Focke-Wulf were devastated in conjunction with Royal Air Force Bomber Command night-time raids. The heaviest raid on Bremen of the war was 18–19 August 1944. In just 34 minutes 274 aircraft dropped 1,120 tons of bombs. The last combined Eighth Air Force and Royal Air Force raid on Bremen was on 22 April 1945.

The 100th's contribution to the destruction of Bremen's industrial–military complex was substantial. The losses in October 1943 were terrible. General Curtis LeMay, who had just put on his fifth star, decided to take the 100th off operations while new crews acquired more time training on the B-17 over the English countryside. Once new planes and crews arrived to replace the 100th's casualties they were initially tucked back in B-17 formations while they gained combat experience. The 100th Group navigator, Major Harry H. Crosby, had his work cut out to integrate these newly trained and limited-experience crews into their first operations. The accounts of his tireless efforts and courage are in his wonderful book, *A Wing and a Prayer: The Bloody 100th Bomb Group of the US Eighth Air Force.*

The 100th was selected for 'Pathfinder' missions. Crews were trained at Alconbury, near Huntingdon. In August 1942 control of RAF Alconbury passed to the Eighth Air Force, flying both the B-17 and the B-24. It was occupied by the 92nd, 93rd and 95th bomb groups, with four squadrons in each of these groups. King George VI visited the USAF base on 13 November 1942 and saw the B-17s based there.

The 100th Pathfinders were trained on the new 'MICKEY' radar system. For its day this was revolutionary, akin to the Norden bombsight in terms of its effects on providing accurate targeting by the Eighth Air Force of strategic targets.

The Norden was an early tachometric system that directly measured the B-17's ground speed and direction, which older bombsights could only estimate with lengthy manual procedures. It used an analogue computer that continuously recalculated the bomb's impact point based on the changing flight conditions, plus an autopilot that reacted quickly and accurately to changes in wind and other effects. (The *Enola Gay* used the Norden bombsight in its operation from Tinian in dropping the Little Boy atomic bomb.) The 100th therefore enjoyed unprecedented accuracy for daylight bombing from high altitudes. The circular error probable (CEP) was 75ft when tested before the system went operational with the Eighth Air Force. The existence of the system was kept secret well into the war. It enabled the B-17s equipped with turbochargers on their four Wright Cyclone 9 radial engines to fly at higher altitudes, increasing the range of B-17 formations.

General LeMay instigated several reforms within the Eighth to improve the use of the Norden sight, including introducing the 'Combat Box' to provide maximum defensive firepower by densely packing formations against air attack. LeMay also had the best bombardiers identified and ensured that they were in the lead bomber in each box. The 100th lead bombardiers did exceptional work combining the Norden sight with the MICKEY radar. The latter enabled 100th bombardiers to see through the clouds to find the target.

These systems made huge differences. However, in the last months of 1943 the 100th lost many crews. The 100th operation on 7 November 1943 over Bremen was characterised by Group Navigator Harry Crosby as 'A real botcheroo'. He flew his eighteenth mission on 13 December 1943.

The Kriegsmarine U-boat base 'Killian' in Kiel was designed as a shelter for newly built U-boats. Nearby was the Howaldtswerke submarine building yard. These facilities were under constant attack by both RAF Bomber Command and the Eighth Air Force. On 4 January 1944, the 100th Bomb Group launched a major attack on Kiel. This attack was Group Navigator Crosby's twentieth mission in B-17s. Very sadly Crosby reflected that by year's end 1943 all the people with whom he had trained were now dead. Overall, the US Eighth Air Force losses were higher than any other US military service. In 1943–44 they lost more men than the entire US Navy in the same period. These losses were not in vain. For example, the 100th flew to Norway to attack the heavy water plants that were being used for Hitler's atomic bomb project.

The period 4–6 March was particularly bad for the 100th and the Eighth Air Force overall. They attacked Berlin. At the conclusion of operations, the Eighth had lost sixty-nine B-17s, fifteen from the 100th.

Within the Eighth overall, 349 aircraft were damaged so badly that they could not fly the next day. Within the 100th morale remained high, and this was a direct result of persistently good leadership by successive group commanders. Two notable commanders were John M. Bennett and Tom Jeffrey. Crosby made this observation about them:

> As I saw it after my study at Oxford, Bennett and Jeffrey had changed the 100th from its original hot fly-boy individuals to twentieth-century work-together warfare. From Romanticism to Neo-Classicism. History in the making. (*A Wing and a Prayer*, p.204).

These group commanders had to face overwhelming challenges. Take for example the navigator community of the 100th guiding the B-17s to targets and back from Norway in the north, to the Bay of Biscay German submarine bases in the west. There were thirty-five original 100th navigators that arrived in England. Only five of these got through to see the conclusion of their missions. The remainder were either KIA (Killed in Action), MIA (Missing in Action) or PoW (Prisoner of War), or wounded so badly that they had to be sent back to the United States. Fourteen per cent of 100 Bomb Group navigators got through, an 86 per cent loss.

It took extraordinary leadership to keep morale high and sustain what became more and more demanding operations. Courage was mixed with greater and greater resiliency based on outstanding training improvements.

The heaviest 100th Bomb Group losses that earned it the sobriquet 'The Bloody 100th' were as follows:

17 August 1943: Regensburg 9 crews lost.
8 October 1943: Bremen 7 crews lost.
10 October 1943: Münster 12 crews lost.
6 March 1944: Berlin 15 crews lost.
24 May 1944: Berlin 9 crews lost.
31 July 1944: Merseburg 8 crews lost.
11 September 1944: Ruhland 12 crews lost.
31 December 1944: Hamburg 12 crews lost.

In between these major losses were others. It took special qualities within each squadron of the 100th as well at the Group Commander level to sustain what was the highly effective strategic daylight bombing that brought German war production to its knees and destroyed the bases from which the Luftwaffe and the Kriegsmarine could continue to attack. The strategic goals were attained, culminating in the support that the 100th gave in preparation for the D-Day landings on 6 June 1944, and the subsequent support for Allied ground operations as they advanced into Germany and the collapse and surrender of the Third Reich.

Before D-Day the 100th had two groups, A and B, flying forty-two aircraft with twelve supernumerary aircraft in support. P-51 Mustangs flew with the 100th B-17s. They attacked strategic targets, factories and communications, and tactical targets, bridges, railways, and airfields. Once the Allies landed, the 100th provided direct tactical support to ground forces. The role of the 100th in ground support intensified as the Allies moved east from the beachheads.

For example, on 24 July 1944, two corps of the US First Army, the Fifth and Nineteenth, were face to face with the German Army. The Eighth Air Force, with the 100th Bomb Group as a key attacking component, literally blasted a corridor 4,000 yards wide in the German lines south-east of Caen. This operational trend continued through 1944 and into 1945 and the war's end.

On 22 February 1945, General Carl A. Spaatz issued an historic order to General Jimmy Doolittle. General Spaatz was the overall Commander of the US Eighth Air Force. He later became the first Chief of Staff of the newly formed United States Air Force in 1947. 'Tooey' Spaatz, as he was known by his nickname, became Commander of the Eighth Air Force in May 1942, setting up his headquarters in England in July. Lieutenant General Jimmy Doolittle, as operational head of the Eighth Air Force (General Spaatz, as overall Commander of US Strategic Air Forces) reported directly to General Eisenhower. He ran the day-to-day Eighth Air Force operations. Spaatz directed grand strategy and, with the RAF's Air Chief Marshal Tedder, agreed that the early key strategic targets were German oil and railway targets, thereby destroying the Germans' ability to fuel and move its war machine. Later, after Victory in Europe Day (VE Day) Spaatz moved to the Pacific to take overall command of US Air Forces, with his headquarters in Guam, from which he directed the strategic bombing of Japan, including the atomic bomb attacks on Hiroshima and Nagasaki. He was personally present at the German surrenders at Reims and Berlin, and aboard the battleship *Missouri* in Tokyo Bay when the Japanese surrendered on 2 September 1945. He was the only man of full general rank or equivalent present at all three of these surrenders.

The 22 February 1945 order from Spaatz to Doolittle stated that the Eighth Air Force was to change its primary strategic bombing mission to tactical. The intent was very clear, to support US ground troops as they fought their way across Germany. He directed that their main targets should be German rail and highway communications, thus disrupting German ability to sustain operations. In addition, there was another key tactical target.

The German 'jet threat' was known from good intelligence sources. The new German jets flew at 540mph, almost 200mph faster than the P-51 Mustangs, the Eighth Air Force's 'Little Friends'. The 100th together with other Eighth Air Force groups, plus the Ninth Air Force fighter aircraft, attacked the airfields where the German jets were based. This was the only effective means to destroy this new and deadly German capability. In due course the 100th helped strafe and bomb every German jet base known to Allied intelligence. They succeeded, denuding German capabilities by the war's end. Yet another triumph and contribution by the 100th.

As the war was nearing its end, the United States beloved President and wartime leader, Franklin Roosevelt, died on 12 April 1945. The 100th mourned his great loss, and determined under the new President, Harry Truman, to fulfil the late President's goal of the final surrender of Germany and the destruction of the Third Reich.

It is salutary to reflect not only on the huge sacrifices made by the brave men of the 100th, and the Eighth Air Force overall, but also to recognise how young these men were when so many made the ultimate sacrifice for bringing Germany to surrender. Most were in their early twenties. Harry Crosby, as 100th Bomb Group's navigator, felt old when he was 26 on 18 April 1945. He felt that he was 'an old man' compared with all the other men in the 100th. These men gave their all. By the time of the German surrender the 100th had lost 732 men killed in action, while 1,040 had been shot down and were prisoners of war.

There is a very appropriate quotation in *American Heritage* by Carl Sulzberger that encapsulates the primary strategic bombing achievements of the 100th. It is an observation by a German Luftwaffe officer, an enemy of the 100th:

> There is no doubt that the Americans harmed us most. The Russians were negligible as far as the home front was concerned, and we could have stood the British attacks on our cities. But the American devastation of our airfields, factories, and oil depots made it impossible for us to keep going.

These words from the enemy sum up the major effectiveness of the 100th Bomb Group's contribution to the overall Eighth Air Force's role in the final surrender of Nazi Germany.

It is important today to reflect on the relevance of the 100th and the Eighth Air Force in the current and future strategic context. Earlier, the roles and missions of the United States Air Force post-Second World War in the United Kingdom and Europe as a whole were briefly alluded to in the context of the significant basing in the UK of United States Air Force capabilities. These have been, still are, and will be retained into the foreseeable future. The United States air assets and the intelligence systems and organisation that support these capabilities are prodigious.

There is a critical strategic issue within the above observations. During the war the British and United States air leadership did not always see eye to eye, although generally there was harmony between the Eighth Air Force and Royal Air Force Bomber Command. However, there was some disharmony between, for example, General Spaatz himself and the Royal Air Force Air Chief Marshal Sir Trafford Leigh-Mallory, who died on 14 November 1944 in an aircraft accident in the French Alps. His position at the time was Commander of Allied Expeditionary Air Force. He had been the Air Officer Commanding (AOC) 12 Group Royal Air Force during the Battle of Britain and had serious differences with the Group Commander who is recognised as the key RAF leader that won the Battle of Britain, Air Chief Marshal Sir Keith Park. He quarrelled with Park over major tactics. Park was right and Leigh-Mallory was proven to be wrong. In the great scheme of things none of the disagreements between Leigh-Mallory and Spaatz made any material difference. Overall, relationships between the Eighth Air Force and the Royal Air Force were excellent.

What are significant are the strategic policy issues that underpinned the whole air war over Europe and the Atlantic. In 2022 it was the 100th anniversary of the death of the British naval strategist Sir Julian Corbett (1854–1922). His *Principles of Maritime Strategy* are studied today in close association with the work of the American naval strategist Alfred Thayer Mahan. Corbett put naval strategy in its historical context. He concentrated on what naval warfare meant to the power of nations and their vital national interests. He demonstrated significant differences between land warfare and war at sea. He concentrated not on the physical destruction of the enemy, rather the actual control and passage on the sea. Control of the lines of communication at sea was vital in his theory. Today naval strategists would call this 'control of the sea'. Corbett rejected the principle of concentration of force at sea. Of relevance for this discussion is Corbett's key point about the furtherance of the national interest using maritime power. In addition, he advocated the diplomatic alliance system and coalitions formed before and after a war. After the First World War and before he died in 1922, Corbett was critical of the British land strategy in the 'war to end all wars' in which so many British lives were needlessly wasted on the Western Front. It was his opinion that the United Kingdom should have pursued a maritime strategy to defeat the German Navy, protect shipping against the new U-boat threat and challenge Germany overseas by the use of its considerable naval power.

The above fits in with Lord Patrick Maynard Stuart Blackett's view that the United Kingdom should have concentrated its efforts and resources on a maritime strategy in the Second World War, and particularly the anti-U-boat campaign, and not used valuable resources on Bomber Command's night-time offensive over Germany. The Eighth Air Force came to the UK from 1942 onwards, after Pearl Harbor and the US Declaration of War against Germany and Japan. The US Air Force has stayed in the UK to this day, with significant capabilities. There is a critical message in the Epilogue that both Julian Corbett drew about maritime strategy and what Lord Blackett derived from extensive Second World War operations research based on excellent science.

What they concluded persisted at the strategic level through the post-war period, into the Cold War and now endures into the post-Cold War and the current world situation with threats to the Western democratic tradition from China and Russia. The future rests on how this process of strategic continuity is perceived and interpreted at the highest levels of the United Kingdom's government. This will be explored further in the latter part of the book.

Meanwhile, the 100th Bomb Group will be examined in microcosm through the experiences of one navigator in one squadron.

# Chapter 3

# One Man, One Navigator of the 100th Bomb Group Paul 'Fred' Julian, 351st Squadron

The Boeing B-17 Flying Fortress was the iconic four-engined bomber of the Second World War, together with the Royal Air Force's Avro Lancaster. The latter made its first flight on 9 January 1941, and was introduced into RAF service in February 1942. Avro built 7,377 of the type.

The B-17 that Paul 'Fred' Julian navigated had a different history to the Lancaster.

The original US Army Air Corps specification for a four-engined bomber that became the B-17 was issued in 1934. It was far-sighted in its detail. It called for a long-range aircraft that could perform daylight precision bombing at high altitude, such that it could avoid the known and projected range of anti-aircraft fire. This required turbo-charged

*Quittin' Time*, a great B-17. (© David Julian)

engines. In due course it would be fitted with the Norden bomb sight. Small-scale production began in 1937.

The B-17 had a crew of ten: a pilot, co-pilot, a navigator-radioman, and six gunners. The six gunners were in the upper fuselage, the belly of the B-17, and the tail. Each gun position was power-operated except for the tail gunner's position. Each gunner operated two .50 calibre machine guns. The later 'G' version of the B-17 entered service in 1943 at the time when the 100th Bomb Group flew to the United Kingdom, with earlier Eighth Air Force command structure and groups arriving in 1942. The 'G' became the definitive version of the Eighth Air Force, with thirteen .50 calibre guns, including a 'chin' turret for defence against head-on attackers. Boeing produced 12,730 Flying Fortresses together with the Lockheed, Douglas, and Vega aircraft companies.

The B-17 operated typically at 25,000–35,000ft, with a maximum bomb load of 8,000lb. On a long-range bombing mission, Eighth Air Force groups would typically carry 4,000lb of bombs. By the end of the Second World War the mighty Flying Fortress was superseded by the B-29 Super Fortress, and after the war ended the B-17s retained in service were used by the new United States Air Force (formed in 1947) for search and rescue missions.

The B-17s of the 100th Bomb Group normally flew in a tight formation for defensive reasons and to concentrate their attacks over the target area. These formations were termed 'boxes'. A 100th box consisted of nine or twelve aircraft. The 100th could put up all its squadrons in one mighty box. When the Eighth Air Force command put up three groups this extraordinarily powerful force made up what was termed a combat wing.

The 100th Bomb Group's 351st Squadron that Paul F. Julian joined as a trained navigator was therefore well equipped.

The Royal Air Force did order B-17s. The RAF's Sir Hugh Dowding of Battle of Britain fame visited Boeing and met senior executives led by C.L. Egtvedt. Later, the US Army Air Corps sent twenty B-17Cs to the RAF and nineteen B-17Fs were sent to RAF Coastal Command for use in anti-submarine operations. The chief of the B-17 design team, Edward C. Wells, was a genius in his own right and in retrospect it would have much benefitted the British and the RAF to have listened to him. Wells subsequently led the development of the follow-ons to the B-17, the B-24 and the B-29. The US Air Force never did make any form of comparisons between the B-17 and the Lancaster, sufficient to say that General Carl A. Spaatz made this highly important observation: 'Without the B-17 we might have lost the war.'

If there was one weakness of the B-17 it was that it could not successfully bomb moving targets. For example, at the Battle of Midway, a critical turning point in the war in the Pacific, high-altitude B-17s were not successful in hitting moving Japanese ship targets, including the aircraft carriers. US Navy torpedo aircraft did the job successfully.

Relations between the US Army Air Force and the Eighth Air Force and the Royal Air Force began early on. In 1941 then Colonel Ira C. Eaker was sent as a 'special observer' to the RAF. In 1942, after Pearl Harbor General Arnold, the overall commander of the US Army Air Force, sent now Brigadier General Eaker to set up the US Eighth Air Force operations in the UK. Arnold sent Eaker this key message: 'To make the necessary preparations to insure competent and aggressive command and direction of our bomber units in England.'

Eaker proceeded to set up the Eighth Air Force at Pinetree in High Wycombe, a girls' school before the war. He organised the Eighth into bomb groups: four squadrons of twelve B-17s in each group, with typically forty-eight B-17s in each of the Eighth's bomb groups. Eaker soon became aware of British weather. Fog became a serious problem for the Eighth when joining up in the air prior to flying missions to Europe. In early training and later when flying missions there were several mid-air fatalities.

Between 14 and 23 January 1943, at the Allied Casablanca Conference a major decision was made that impacted the Eighth's bomb groups. It was at this gathering that General Spaatz made the case in front of the Allied leaders for daylight strategic bombing, at variance with the views of Winston Churchill. The British Prime Minister wanted the Eighth Air Force to combine with the RAF in night-time bombing. President Roosevelt, and generals Spaatz and Eaker, said a resounding 'no'. The Americans stuck to their guns, literally and metaphorically, and won the day, a critical war-winning decision.

The 100th flew its first mission on 25 June 1943. The 'Bloody 100th' title was conceived after 21 aircraft with over 200 men went missing or were killed in one week in early October 1943. The 100th flew a total of 306 missions, 8,630 sorties, and released 19,257 tons of bombs and 137.8 tons of supplies. Over 800 men from the 100th were killed in action or flying accidents or became prisoners of war. A total of 229 B-17s from the 100th were either lost or salvaged. At any one time 3,500 men served at the Thorpe Abbotts base (data from *Century Bombers: The Story of the Bloody Hundredth*, by Richard Le Strange).

Other bomb groups of the Eighth Air Force had begun operations before the 100th arrived in England. For example, in June 1943 the *Memphis Belle* crew became the first Eighth Air Force aircraft to complete its mission quota of twenty-five missions with its crew still intact. The 100th was not so fortunate. On 6 March 1944 it lost fifteen aircraft over Berlin. On 8 March, Lieutenant Colonel John M. Bennett led the 100th in a courageous attack on Berlin, winning the Silver Star.

Generals Doolittle and LeMay attended the 100th's '200 Mission' party. Sadly, other events clouded such celebrations. For example, Major Glenn Miller, who led the US Army Air Force Band, left from the 100th Bomb Group on 15 December 1944 for France and his aircraft disappeared over the English Channel. It is not known what happened. It was thought that perhaps his aircraft was hit by a returning B-17 disposing of its unused bomb load over the Channel, although there was no hard evidence for this.

The B-17 aircraft that Paul. F. Julian flew in as navigator of 351st Squadron was a standard Boeing aircraft just like all the other B-17s in the Eighth Air Force. The B-17 would later be replaced by the B-29, which had fully retractable landing gear, a pressurised cabin, and remotely controlled gun turrets. In the Pacific theatre the B-29s would fly 3,000 miles sorties from Guam in the Mariana Islands to attack Japan.

By the end of the Second World War, the 100th had been awarded two Distinguished Unit Citations, for the 17 August 1943 attack on Regensburg, and for the 4, 6 and 8 March 1944 attacks on Berlin.

The Eighth's strategic bombing campaign brought German industrial capacity and capabilities to its knees. This campaign enabled US, UK and Canadian forces to take the fight into Europe after the successful D-Day landings that led to final victory and liberation from tyranny. It is to the memory of all the men who sustained this colossal effort, and the sacrifices made that today's relationships between the United States Air Force and the United Kingdom endure.

Photographs are worth a thousand words. The following photographs show what it was like for navigator Paul 'Fred' Julian of the 351st Squadron of the 100th. Fred's experiences are encapsulated in the timeless book written by the 100th's group navigator, mentioned earlier, Harry Crosby. His *A Wing and a Prayer* cannot be replicated and it would be totally inappropriate to do so. Harry Crosby consulted Fred Julian while writing his book, so what is in *A Wing and a Prayer* says it all and captures what the brave men of B-17 *Quittin' Time* endured.

The following is a reflection of Harry Cosby on Fred Julian, the archetypal navigator of the 100th:

> Freddie was all business on the base at Thorpe Abbots and in the air. He would wake up early and check his 50-cal gun even if the crew had already done so. He would also check the maps, his tools and whatever he was responsible for in the ship. During missions, his job was so crucial that during AA and fighter fire, they had to yell at him to drop his slide rule and take cover.

What more can be said? It would be churlish to even try to elaborate.

Fred Julian survived and returned to the United States to become a business owner, wonderful husband and father of four. Since Fred's passing in 1992, his son David, honours his father's illustrious memory and service during the war.

As with so many veterans, Fred rarely revealed details of his service and the harrowing missions. When Fred passed in 1992, David discovered an old envelope of photographic negatives in a file cabinet drawer. Once printed, these amazing images rekindled a curiosity to learn the untold stories of his father's life, and to more fully understand what Fred and his crew experienced during the war. From his research, David discovered that *Quittin' Time*'s co-pilot Walter Scott lived only a few miles from him in Brooklyn. David got in touch and spent a memorable afternoon with him and his wife. Walter filled in many details of life during *Quittin' Time*'s missions.

One story that was always recalled about *Quittin' Time* arrived in a letter from bombardier John Frascatore about tail gunner Doug Wright (who well-earned the moniker 'Dead-Eye Doug'), and his interactions with attacking German fighter pilots. Of special note is the story of the harrowing day on one Berlin mission when a different B-17 borrowed by the crew while *Quittin' Time* was needing repair was 'shot to hell'.

This was why Fred Julian's *Quittin' Time* was considered a 'lucky plane' by the rest of the Bloody 100th. After Fred and his fellow crewmen completed their missions, the data suggests that their aircraft *Quittin' Time* flew the most missions in the shortest time.

For readers who wish to delve further, do visit the 100th Bomb Group Museum near Diss in Norfolk and consult the fine work that is still done by Mike Faley, who has been devoted for years to the history of the 100th and was instrumental in helping David learn and share more about *Quittin' Time* during Fred Julian's service. Or visit the Mighty Eighth Museum in Savannah, Georgia. Also consult the fine work and

Damaged B-17 of Fred Julian's squadron, the 351st. (© David Julian)

archives found online by the 100th Bomb Group historical organisation (www.100thBG.org).

The following photographs were assembled by Fred Julian's son, David. Supporting commentary was provided by Mike Faley.

As reported by *Quittin' Time* Bombardier John Frascatore:

> Shown here is the result of an anti-aircraft shell that tore through the radio room and went through the ship before exploding. It left behind a gunner who had just begun to move to the other side when he felt the rush of air on his face as the shell tore through. It also took with it the entire right gun position. All that was left was a mess of exposed hanging control cables. After landing we counted over 250 holes in the plane before we got tired. The miracle was that no one was injured!

B-17 serial 42-31530 EP-X was nicknamed *Quittin' Time* because the last three digits, 530, signified quitting time at work, 5.30 pm.

*Quittin' Time* (centre plane) in formation piloted by Ernest T. 'Tracy' Gilbert. (© David Julian)

Fred Julian's flying jacket, 351st Squadron, 100th Bomb Group. (© David Julian)

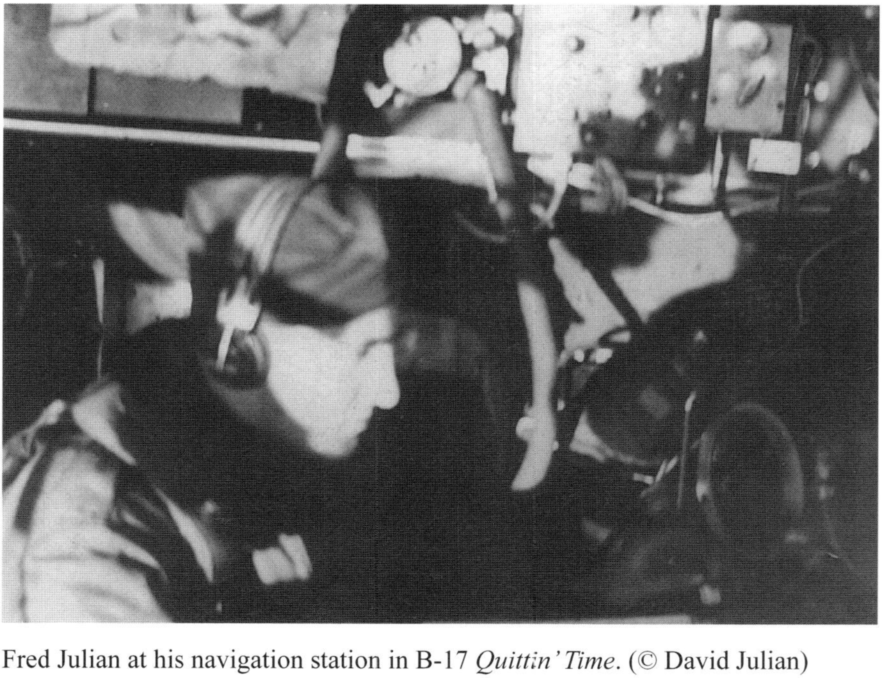

Fred Julian at his navigation station in B-17 *Quittin' Time*. (© David Julian)

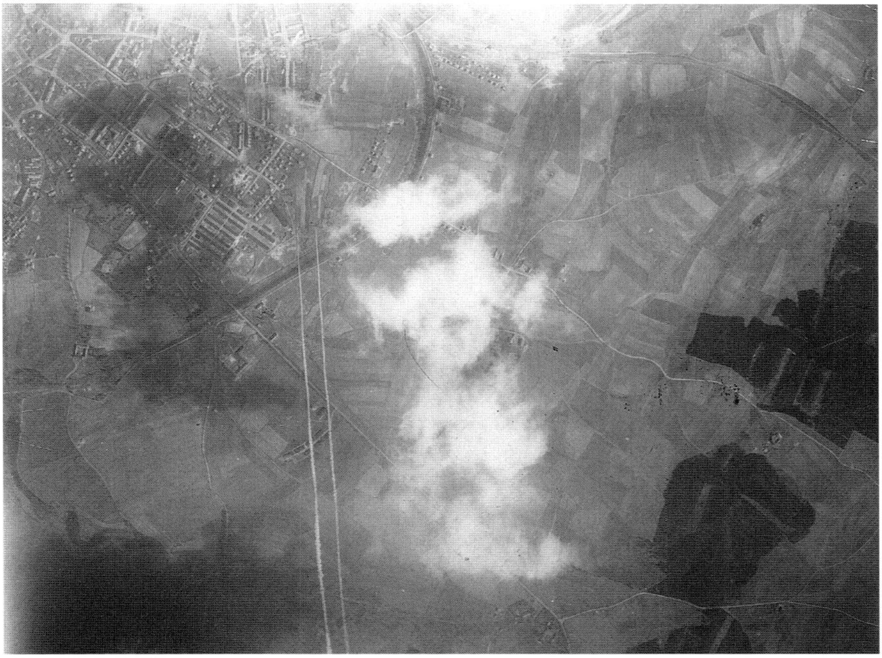

View from B-17 *Quittin' Time* on the Eger mission. (© David Julian)

B-17 *Quittin' Time* statistics, with ten-man mission crew as well as ground crew members of the 351st. (© David Julian)

B-17 *Quittin' Time* bombardier John Frascatore approaching the mission's target. (© David Julian)

*Quittin' Time*'s mission crew joined the 100th Group on 24 December 1944. They were: 1st Lt Ernest T. Gilbert (pilot), 1st Lt Walter G. Scott (co-pilot), 1st Lt Paul F. Julian (navigator), 1st Lt John R. Frascatore (bombardier), Cpl William Runge (radio operator), Cpl Donald V. Gloege, Cpl Kenneth L. Morphew (top turret gunner), Cpl George G. Elliott (ball turret gunner), Cpl William G. Tunnessen (waist gunner), Cpl Douglas N. Wright (tail gunner).

Cpl Doug Wright shot down an Me 262 jet on 18 March 1945, over Berlin.

Strategic bombing missions flown by *Quittin' Time*:

01. 14/1/45 DERBEN
02. 17/1/45 HAMBURG
03. 20/1/45 HEIBRON
04. 21/1/45 MANNHEIM
05. 28/1/45 DUISBURG
06. 29/1/45 KASSEL

07. 3/2/45 BERLIN
08. 6/2/45 BÖHLEN
09. 15/2/45 COTTBUS
10. 17/2/45 GIESSEN
11. 19/2/45 MÜNSTER
12. 21/2/45 NÜRNBURG
13. 22/2/45 WEHINGER
14. 23/2/45 TREUCHTLINGEN
15. 25/2/45 MUNICH
16. 10/3/45 DORTMUND
17. 11/3/45 HAMBURG
18. 12/3/45 SWINEMÜNDE
19. 15/3/45 ORANIENBURG
20. 18/3/45 BERLIN
21. 21/3/45 PLAUEN
22. 22/3/45 ALHORN
23. 23/3/45 UNNA
24. 24/3/45 STEENWIJK
25. 24/3/45 ZIEGENHAIN
26. 28/3/45 HANOVER
27. 30/3/45 HAMBURG
28. 31/3/45 ZEITZ
29. 3/4/45 KIEL
30. 4/4/45 KIEL
31. 5/4/45 NÜRNBURG
32. 6/4/45 LEIPZIG
33. 7/4/45 BUCHEN
34. 8/4/45 EGER

Notes on a typical mission by Lieutenant Frascatore:

18 March 1944, Target: BERLIN.

'We were flying auxiliary lead below and behind the lead ship in the low squad. We were suddenly jumped by enemy planes – four or more jets included. In no time at all, we lost our leader as well as 3 other planes (Swain, Jensen, King, Gwin). A jet was on our tail trying to finish us off but fortunately our Tail Gunner, Doug Wright, shot him down instead. Naturally, our low formation as well as a good part of the group had scattered badly. At that point, our pilot Tracey Gilbert assumed leadership and flew straight & level permitting the others to form on our plane. The planes soon re-grouped, and no further losses were experienced. That evening Col. Harry Cruver recommended Gilbert for a promotion to Captain. We officers had just been 1st Lts for 2 weeks, so Tracey was a 1st Lt for a very short time.'

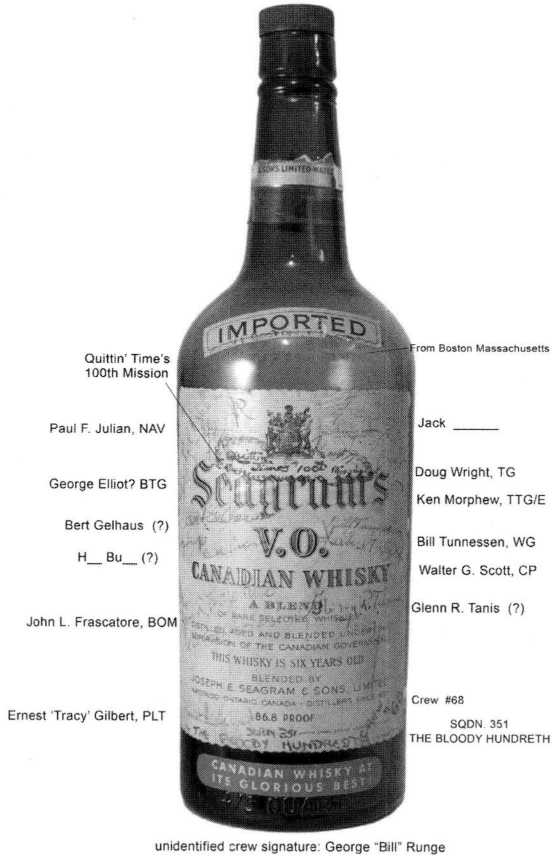


The crew of *Quittin' Time* led by Tracey Gilbert survived 34 of the 100 missions completed by this aircraft in the Eighth Air Force. The bottle of Seagram's whisky was consumed by the crew to celebrate the ship's 100th mission. Inside the bottle is a 3in piece of shrapnel recovered from the collar of a crewmember's flight suit after a particularly harrowing bombing mission to destroy German aviation supply factories. (© David Julian)

The following is the transcript of a poignant letter from Fred Julian's son David to one of his father's fellow crewmen:

> 11/23/92
>
> Dear Mr Brown,
>
> My name is David Julian, son of Fred Julian of the 100th bomb group. I just came across your letter dated 10/12/92 sent to my father. The family is all here for the Thanksgiving holiday, but I am sorry to say that Dad passed away suddenly just the other night (11/21/92) after watching his favourite game, the Harvard versus Yale football game and then eating a wonderful dinner. Dad had a great day, was in good spirits and did not suffer when his heart failed. We bury him tomorrow morning and have made sure of a mention of his proud service in the 100th group, and his ongoing interest in his fellow squadron members. I have always prized the photos of the crew of the *Quittin' Time* in front of that massive B-17, so one of them will always be mounted above my desk as an inspirational memento of bravery and camaraderie.
>
> I deeply appreciate your involvement in recording and maintaining the history of the 100th group, so I wanted you to know of Fred's passing. I also hope that you are doing well since your surgeries, and our family sends you our best wishes. If you ever wish to get in touch, or require our help concerning Fred's background or photos, write to his wife with the same address that you have, or to me.
>
> Sincerely,
> David Julian.

In the third and final part of this book, we will examine the current and future regular deployment of major United States Air Force assets and capabilities to the United Kingdom. There has been continuity from 1942, when the Eighth Air Force arrived with the B-17, to today with the availability of the B-2 Spirit, the B-52, and into the foreseeable future with the B-21.

The B-2 Spirit.

The strategic bomber significance today and in the future is one cornerstone of US-UK defence and security relations based on forward-deployed deterrence capabilities and operations.

Today USAF advanced aircraft systems in the UK provide the British Ministry of Defence with the opportunity to concentrate resources into the strategic maritime domain. The latter enables the UK to work closely with the US Navy and Marine Corps in forward-deployed deterrence operations dedicated to the preservation of peace at a time of serious threats to global stability. Maritime power in the pursuit of peace provides the UK with a major role.

The continuity from the great days of the Eighth Air Force and the brave squadrons of the 100th, and of our strike carriers to the present and into the future is a testament to each individual who played such a crucial role.

Fred Julian, a great 100th navigator, and his fine fellow crewmen, exemplify all that is good in these incredible past exploits that provide such extraordinary examples to encourage us all to go ‘onwards and upwards’ together, the United States and the United Kingdom, and their close allies.

# Part II

# The Falklands Air War

**The history of the Falklands Air War is related in the first person by Commander N.D. MacCartan-Ward DSC AFC, better known as 'Sharkey' Ward.**

**His first book, *Sea Harrier over the Falklands*, now a military classic, was described by Admiral Sir John Woodward GBE KCB, Commander of the Carrier Battle Group, as 'a fully accurate account of the Falklands Air War – save where you presumed to know what was going on in my mind'.**

# Chapter 4

# The Preamble to the War – The Day the RAF Lied

Within the UK, the evolution of strategic air power, as so well described in previous chapters, has been tainted by internecine struggles for budgetary resources between the three services. Critically, historical military precedent has been ignored by the UK's Ministry of Defence and government. On too many occasions, blatant and subjective misguided propaganda and advice has been fed to ministers. This has led to a dangerous lack of objectivity and detailed analysis concerning the latent threats to our island nation's prosperity and survival.

When the Wright brothers showed the world that mankind could take to the air like birds of prey, a euphoria concerning air power and its military application gripped the world. A new and devastating weapon system had been born but, as the Second World War demonstrated, its initial custodians within the UK did not always utilise it to best effect.

Instead of following the wise advice of the Lord Trenchard that the Air Force should be subsumed into the effective war-fighting services of the Royal Navy and the Army, a new empire was born. The then Air Ministry did not have the mental acuity or desire to understand Winston Churchill's famous quote:

> Whereas any European power has to support a vast army first of all, we in this fortunate, happy island, relieved by our insular position of a double burden, may turn our undivided efforts and attention to the Fleet. Why should we sacrifice a game in which we are sure to win to play a game in which we are bound to lose?

No attention was given in the inter-war years to providing our Fleet with state-of-the-art embarked air power. Our Whitehall mandarins ignored the farsighted and rapid build-up of aircraft carriers and embarked air power taking place in the USA and in Japan and this was to cost UK dear in the Battle of the Atlantic, the northern convoys, the Malta convoys and later, the Far East.

As the Second World War wore on, the UK successfully played catch-up with the rapid build-up of its aircraft carrier fleet and significant battle honours were won. These included the sinking of the *Bismarck*, the sinking of the Italian fleet at Taranto and the virtual destruction of Japan's oil supplies at Palembang, Indonesia, that were so vital to Japan's war effort.

Land-based tactical airpower could not and did not answer the call.

What has all this got to do with the Falklands War, 1982? A great deal is the answer.

Post-Second World War and with the NATO nuclear umbrella deterring any thoughts of World War III (as it continues to do today), the size of our armed forces was reduced but the UK still retained a robust Fleet that included the availability of effective sea-based tactical aviation – when embarked this provided a strategically mobile capability. British expeditionary force operations projected strategic maritime airpower to deter our enemies and assist our allies. These were enabled by the existence and deployment of our aircraft carriers – mostly when our land-based tactical air force was unable to take part.

Operations as described by naval historian David Hobbs in February 2011 to the author, included:

## Palestine 1948

Naval aircraft from HMS *Ocean* covered the final evacuation of British forces from Palestine. RAF aircraft had already been evacuated.

## Korea 1950–53

HMS *Triumph* joined the USS *Valley Forge* to strike at North Korean targets shortly after North Korea attacked the South in June 1950. The British aircraft carriers *Triumph*, *Theseus*, *Glory* and *Ocean* provided all the UK's tactical strike and fighter operations throughout the three

A Sea Fury (upper) flying in company with a Sea Hawk.

years of the war, flying thousands of effective sorties. The naval Sea Fury (shown above) flown by 'Hogey' Carmichael registered the UK's first and only air-to-air kill of the Russian-built MiG-15 jet fighter.

RAF land-based tactical fighters and fighter-bombers could not and did not take part.

## Suez 1956

This was a combined assault on Egypt by British and French carrier-borne and land-based aircraft. In the British operations, the RN deployed three fixed-wing carriers, *Eagle*, *Albion* and *Bulwark*, plus two helicopter carriers, *Ocean* and *Theseus*. These strike carriers reacted more quickly to calls for offensive air support than RAF aircraft in distant Cyprus and Malta. Despite only having one third of the total British strike fighters, embarked RN aircraft flew two thirds of the strike sorties and their aircraft spent longer over the target area. RAF aircraft had long transits from their bases, carried fewer weapons and could spend little time on task, most of that at high level to conserve fuel (shades of Libya, 2011).

## Levant 1958

US-UK assistance sought to protect Lebanon and (land-locked) Jordan against Iraqi aggression. HMS *Eagle* provided support for airborne and amphibious forces deployed into theatre.

## Korea 1960

UN forces including a RN carrier deployed to the Yellow Sea on exercises aimed at deterring the North from launching a renewed attack on the South. Deterrence succeeded. There was no RAF involvement.

## Kuwait 1961

British forces deployed to Kuwait to defend it against threatened Iraqi aggression. HMS *Bulwark* arrived with 42 Royal Marine Commando within twenty-four hours (since good intelligence had put her in the right place) and used her helicopters to deploy and support them. Strike carrier HMS *Victorious* took several days to arrive with her battle group from the South China Sea but brought the 'complete package of air power' that subsequently dominated the combat theatre.

Aerial view of the Royal Navy aircraft carrier HMS *Victorious* (R38), taken around 1958–60, when the Royal Navy operated the Douglas Skyraider AEW.1.

## Confrontation with Indonesia 1963–66

Britain and the Commonwealth supported the Malaysian government against Indonesian aggression and deployed forces from all three services. The Far East Fleet provided a considerable deterrent against Indonesian escalation and the presence of its strike carriers posed a threat that Indonesia could not counter. Carrier and air group transits of high-visibility international waters such as the Sunda Strait added to their deterrence value. Land-based air could not provide such a visible deterrent.

## East African Mutinies 1964

Following a mutiny by Tanganyika army units in January 1964, Britain was asked to provide assistance. HMS *Centaur* was at Aden and embarked 45 Royal Marine Commando, 16/5 Lancers with their vehicles and two RAF helicopters in addition to her normal air group. The subsequent assault was 'a model of how flexible carriers are and how quickly they can act'. RAF tactical fighters could not and did not take part.

## Defence of Zambia 1965

Following the Unilateral Declaration of Independence by Rhodesia, the Zambian government requested air defence arrangements from the UK. HMS *Eagle* was positioned from 23 November to 7 December 1965 pending the arrival of RAF fighters.

## Beira Patrol 1965–66

Following on from assistance to Zambia above, Britain undertook to enforce UN sanctions preventing tankers from entering Beira with oil for Rhodesia. Only carriers could search the vast areas of sea involved in the months it took the RAF to build up a maritime patrol aircraft base and deploy aircraft to it. *Eagle* and *Ark Royal* were both involved for considerable periods at sea. RAF tactical fighters could not and did not take part.

## Aden 1967

British forces were evacuated from Aden in November 1967 covered by an RN task force offshore. RAF aircraft were among the forces evacuated and relied on RN carrier-borne aircraft for their defence while they did so.

## Belize 1972

A show of strength by Buccaneers from *Ark Royal* prevented a threatened invasion of British Honduras (Belize) by Guatemala. Land-based air was too far away and could do nothing.

There were other occasions when RAF resources were deployed offshore, but these resources had to rely upon the Royal Navy to carry them to their destination.

Although the RAF would have politicians believe that it is a mobile force, there are a number of examples where it could not have deployed aircraft, aircrew and maintainers into conflict zones without aircraft carriers. Once deployed, the RAF then relied on sea-borne bulk supplies of fuel and ammunition that, in turn, needed the RN to maintain control of the sea supply routes with its aircraft carriers playing a prominent role.

Examples include:

Buccaneer bomber.

## Russia 1919

British forces deployed into Russia against Bolsheviks in 1919 needed air support, most of which was transported into theatre in RN aircraft/seaplane carriers and then deployed ashore.

## Palestine 1929

Aircraft from HMS *Courageous* landed to support British troops in action in Palestine.

## Norway 1940

RAF aircraft based in the UK had insufficient range to be effective in operations over Norway after the German invasion. Most support came from carrier-borne aircraft, but the only effective RAF involvement came from Gladiator and Hurricane fighters landed from the aircraft carriers *Glorious* and *Furious*. The carriers also gave navigational and technical support.

## Malta 1940–42

RAF fighters had insufficient range to fly to Malta from Gibraltar after Italy entered the war and the only way to get Hurricanes and Spitfires to Malta was to fly them off the decks of RN and USN aircraft carriers. There would have been no fighters in Malta without them. HMS *Ark Royal* was lost returning from delivering Spitfires to the island.

## North Africa 1941–42

Most of the fighters in the Desert Air Force relied on sea control to get them there and numbers were carried in aircraft carriers such as HMS *Furious*.

## Sumatra 1942

RAF fighters did not have the range to reinforce Allied forces in Southeast Asia as the Japanese advanced in early 1942. HMS *Indomitable* ferried and flew off Hurricanes to Sumatra, which could not, otherwise, have got there.

## Pacific 1945

RN escort carriers ferried RAF Mosquito bombers and their secret 'Highball' bouncing bombs to Australia in 1945. They were intended to operate from the decks of fleet carriers to attack Japanese warships in the home islands, but the USN refused to allow them to do so. The RN put considerable effort into training their crews.

## Korea 1950–53

Although the RAF did not operate combat aircraft in the Korean War, it did supply Meteor fighters to 77 Squadron of the RAAF, which did operate. They were ferried to Iwakuni in Japan in British aircraft

The amazing RAF Mosquito fighter-bomber.

carriers including HMS *Unicorn*. Without her, the RAAF could not have supported the squadron in action.

## Indonesian Confrontation 1963–66

During the Indonesian Confrontation against Malaysia, the RAF relied on RN commando carriers to ferry helicopters to Borneo. They lacked the range to fly there.

## East African Mutiny 1964

Two RAF Belvedere helicopters were embarked in HMS *Centaur*; they had no other means of reaching the scene of action.

## South Atlantic 1982

Initially, RAF Harriers and Chinook helicopters had no means of reaching the conflict other than on the converted containership, SS *Atlantic Conveyor* and the decks of *Hermes* and *Invincible*. They relied on the RN radar/air defence environment and RN supplies of fuel and weapons to be effective, neither of which would have been there without the carriers.

## Iraq 2003

RAF Chinooks embarked in HMS *Ark Royal* took part in the amphibious assault by Royal Marines on the Al Faw peninsula. Again, they lacked the range to operate from a land base and needed the aircraft carrier to take them to the fight.

By the late 1960s, the utility demonstrated by the Fleet's projection of air power, described in brief above, and the inability of the land-based RAF's tactical fighter aircraft to respond to and to carry out nearly all these operations had not been adequately recognised or understood by the UK government. This was a direct result of the poor public relations effort by Their Lordships of the Admiralty and the Naval Staff. Her Majesty's Royal Navy still clinged to the misguided and anachronistic policy of being 'the Silent Service'.

The silence was deafening, and so it was that in 1966–67 came the infamous day that the RAF misled our Secretary of State for Defence, Dennis Healey. He was persuaded by the RAF not to build the planned new aircraft carrier CVA-01 on the extraordinary grounds that: 'The RAF would and could provide air defence for the Fleet throughout the Global Commons – and that therefore we did not need aircraft carriers.'

They went to great lengths to support this false claim, providing the minister with maps of the world that showed how far a single land-based fighter aircraft (they used the performance of the F-111, which was not in the RAF inventory) could reach the distant parts of any ocean. In doing so, they had to create a false map of the Indian Ocean in which Australia was moved more than 500 nautical miles to the west. The minister and his staff did not have the gumption to ask the question, 'What can it do when it gets there?' The answer of course was 'nothing'. It could not loiter over the Fleet and provide 24/7 air defence. Instead, it had to return straight away to its land base – leaving the Fleet defenceless against air attack.

The expeditionary force successes of the previous two decades and the proven lack of utility and deployability of land-based tactical air were summarily forgotten. The Belize Operation in 1972 reinforced the illogicality of the Healey decision – but our government refused to change course.

CVA-01 – artist's impression. (Wikipedia)

It is not unreasonable to suggest therefore that the cancellation of CVA-01 and the subsequent withdrawal from service of our last strike carrier, HMS *Ark Royal*, signalled the demise of the Royal Navy as a global maritime power player and paved the way for Galtieri's decision to invade the Falkland Islands in 1982.

But he didn't count on the Sea Harrier rising from the ashes of the Fleet Air Arm and enabling the recovery of our sovereign Falklands territory by our Royal Marines-led ground forces.

HMS *Ark Royal* at flying stations.

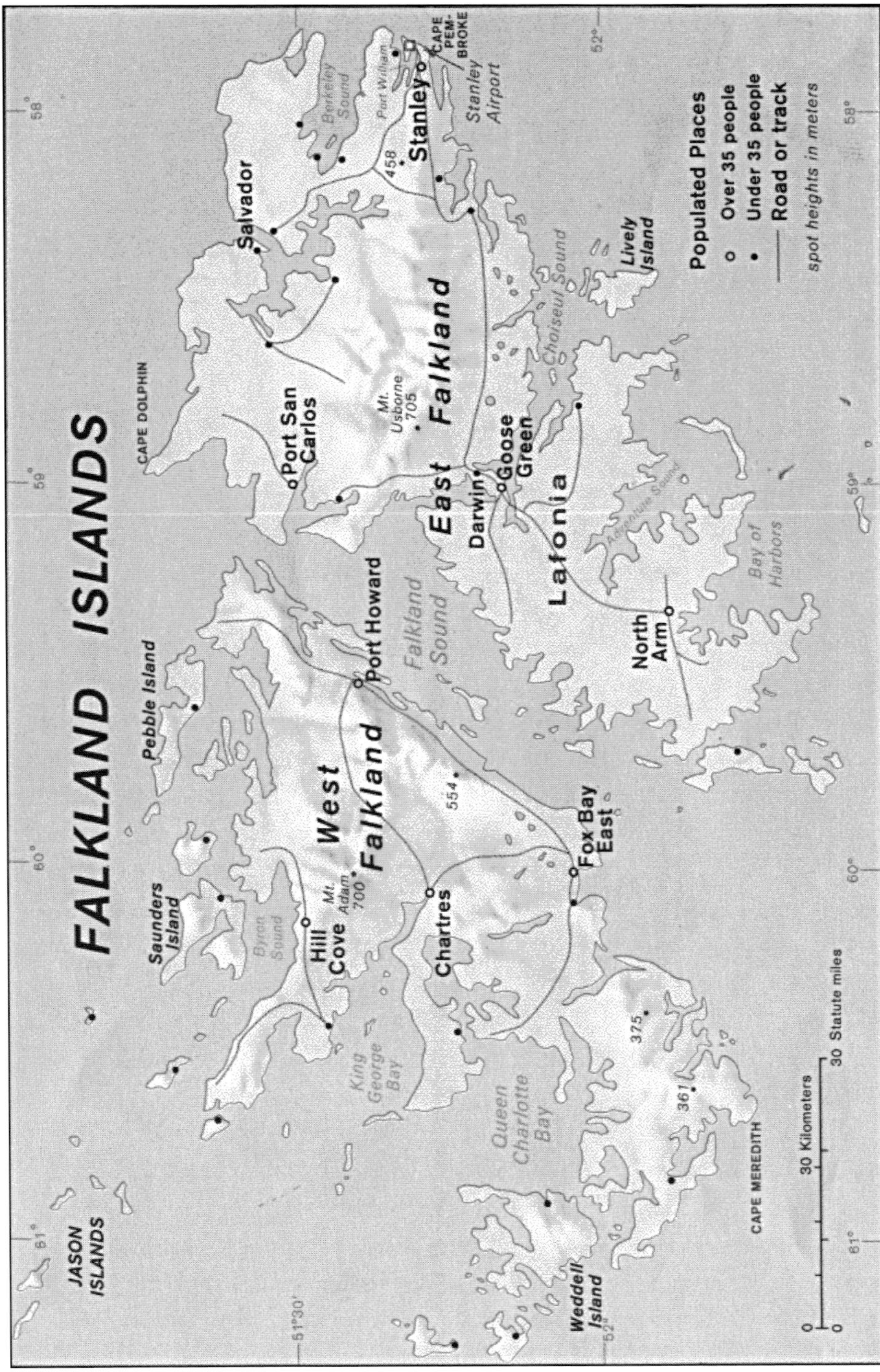


The Falkland Islands.

# Chapter 5

# The CVA-01 Decision in More Detail

The rationale behind the 1966–67 Healey CVA-01 carrier decision merits some detailed analysis. It remains as flawed today as it was then, as evidenced by the continuing misguided criticism of our new carriers by the 'great unwashed' in Whitehall. 'Unwashed' refers to those with a very loud voice who have zero experience of or expertise in naval warfare and, particularly naval air warfare.

In the late 1960s, the Royal Air Force had witnessed two decades of successful Royal Navy expeditionary force operations in which their short-range, land-based tactical fighter aircraft could play no part. Up to that time, the UK's strategic nuclear deterrent had been vested in two delivery options, the long-range bombers of the RAF and the Buccaneer low-level strike bombers of the RN Fleet Air Arm. That strategic nuclear deterrent was now being transferred to the UK's ballistic missile submarines. In other words, the Royal Air Force found itself without any strategic role. The biggest reason for its *raison d'être* had disappeared.

Strategic nuclear bombers continue to carry out a vital NATO deterrent role and these were and still are provided by the United States Air Force, some units of which are regularly deployed to the UK (with less frequency than during the Cold War). This reduces markedly the financial pressures that might otherwise be felt by the UK defence budget.

Couple this with the Cold War status quo regarding the front line of air defence of the United Kingdom against Soviet air attack being comprehensively conducted by the strike carriers of the United States and the Royal Navy, and one can understand the paranoia that had set in within the RAF High Command. How could they justify their continued existence as a war-fighting service?

*Per Ardua ad Panica* should have been their new motto.

Their only option was to mislead ministers about their global capability and to use propaganda to bolster that claimed capability. The phrase, 'the indivisibility of air power' was concocted and rang through the corridors of power in Whitehall. If its meaning had been thoroughly discussed and analysed in the light of recent expeditionary force operations, Whitehall would have understood that it was referring logically to the power projection capability of embarked naval air power. But such analysis was stymied by MoD/Air's persistent and perfidious claim to ministers that, 'We are the Royal Air Force. We are the experts in air warfare. If we say we can do something, we can!'

It was with this facile and contrived background in play that the Secretary of State believed the RAF's extraordinary and misguided claim that, 'UK did not need aircraft carriers. We, the Royal Air Force, can provide air defence of the Fleet throughout the oceans of the world.' Nothing could have been further from the truth, but the decision was made. The last of the UK's conventional strike carriers was to be withdrawn from service in 1977.

It was an appalling decision and the RAF game plan surrounding it should come under the severest scrutiny.

What was going through the minds of the paranoiac junior service? Was their initiative the result of incompetence, ignorance or an overwhelming desire to create, at any national security cost, a new *raison d'être* that depended upon the destruction of the Royal Navy fixed-wing Fleet Air Arm?

For a service that claimed to be the superior authority on all forms of air power, the armchair Air Staff officers who presented the game plan to ministers must have been totally incompetent and blissfully ignorant of their own inability to provide 24/7 air defence of the Fleet throughout the global commons. That is a possibility – but it doesn't really hold water. Instead, they made an unjustifiable promise to ministers that they knowingly could not keep and in doing so were willing to sacrifice the UK's global military power projection capability that, in conjunction with our allies, is so vital to our island nation national security and commercial prosperity.

The manner in which they wilfully misled our Secretary of State and the government has left an indelible stain on their character and their integrity.

# Chapter 6

# Rising from the Ashes – The Sea Harrier is Conceived

I believe it would be fair to say that my personal Falklands experience began more than a decade before 1982. One could say that it was a character-forming period, at least in so far as becoming a true war fighter was concerned. In the beginning I was already a brash, competitive, aggressive, confident and somewhat unruly young officer but I had no illusions about the difficult path that lay ahead – hopefully becoming professional enough to rub shoulders with and then be accepted by those I most admired, i.e., the front-line, fixed-wing fighter pilots of the Royal Navy.

My upbringing in the fold of an RAF family had taught me to respect my elders, seniors and superiors. That was in my nature. But I believed strongly that respect must be earned, particularly within highly professional organisations. Respect does, of course, go hand in hand with loyalty downwards. When that was not the case, I found great difficulty in joining the club, 'Yes sir, no sir, three bags full, sir' – de facto, I never joined it. Loyalty to one's superiors is important in any organisation but there is a critical difference between being, on the one hand, a yes man and, on the other hand, providing one's boss with alternate professional viewpoints. This shaped my journey towards becoming Mister Sea Harrier and taking my team to war. Yes, I have to admit I crossed the dotted line of conventional military obedience on many occasions. But in the end and with the luck of the Irish, that served me well.

In the early 1970s, the fixed-wing Fleet Air Arm was 'down but not out', staring disbandment in the face. But its demise lay in the distant future and was not something I dwelt upon. I was excited and privileged to be appointed directly from operational flying training to join the front-line 892 NAS flying Phantoms on board HMS *Ark Royal*.

The Holy Grail – an 892 Squadron Phantom.

Despite the CVA-01 decision, morale was very high in the ship, and I was surrounded by a superb team of experienced fighter pilot top guns who taught me my trade. They flew hard, played hard and gave no quarter whether on or off duty. It was a formula that suited me completely.

As the new boy on board, I felt I had found the Holy Grail and I knew I had a huge amount to learn from the idols that surrounded me. They all knew that I was a bit of a maverick, having been court martialled very recently for unauthorised low-level flying off the beaches of Cornwall and Devon. But this did not loom large in my mind after making my first deck landing on board on 14 June 1970 – exactly twelve years before the Argentine surrender in Port Stanley. Two days before, the squadron had suffered the loss of Alec Stewart and Pip Coombs on a low-level flight over the Channel. Alec had been the squadron Staff Officer and my boss, Nick Kerr, thought it entirely appropriate that I, as a career officer should assume his duties. I wasn't a very happy bunny – shuffling paper was not my priority as the new boy flying from the deck!

With all the squadrons having landed safely on board, the whole air group of Phantom, Buccaneer, Gannet and helicopter aircrew descended on the wardroom for some convivial social drinking. I was enjoying my beer with a couple of AEW Gannet aircrew at the back of the wardroom when a very loud voice pierced the noisy gathering, 'Staffie, get me a drink!' It was my boss, the ebullient Kerr, who was standing by the bar! The wardroom fell silent, and I became the focus of attention. My Gannet friends implored me, 'Sharkey, that was your boss. You better get moving!' I didn't flinch or move and just carried on with my beer.

'Staffie, get me a f***ing drink!' He sounded furious!

I waited a few milliseconds before responding at the top of my voice, 'Get your own f***ing drink!' My fellow drinkers gasped. All in the wardroom were awaiting the explosion that surely would follow. Nick Kerr, bless him, just grinned and turned away. I had passed the test. It was a good start to a very happy relationship.

Calling a spade a spade is not always a clever thing to do when addressing senior officers. But I believe that all good staff officers should speak the truth, up to the moment that their advice is discounted by their senior officer. That was my *modus operandi* during my first two flying tours with 892 Squadron. My obvious lack of 'officer-like' tact and diplomacy did not hinder me under the command of Nick Kerr and then, Bill Peppé. Under the former as a lieutenant, I became an Air Warfare Instructor and an Instrument Rating Instructor and under the latter, and shortly to be promoted to lieutenant commander, I was appointed as the squadron senior pilot. As the latter, my main contribution was to increase the proficiency of the squadron in within-visual-range fighter combat (dogfighting) and to get rid of the myth that night deck landing was 'a black art'. These two accomplishments were to prove vital when we travelled down south.

On leaving the squadron to go to the Ministry of Defence, one of my old and bold pilots collared me on the flight deck and said, 'Sharkey, you are a crazy bastard but if I ever go to war, I want it to be with you.' There could be no higher praise than that.

In the Naval Staff, my job in the Directorate of Naval Air Warfare (DNAW) was the Sea Harrier desk officer in charge of the final development and production of the Fleet Air Arm's new jump jet fighter. I was in sole charge of the project (whereas the equivalent RAF Harrier desk was supported by a team of fifteen group captains, wing commanders and squadron leaders). I was thrilled to be running the project but although it gave me great job satisfaction professionally speaking, it was a dismal existence working inside the grey corridors of the MoD, especially as my reputation for straight talking had preceded me from the front line. My supposed 'lack of tact and diplomacy' initially did not fit well with my superiors who, without telling me, gave me the worst confidential report in Naval Staff history. But when Captain Lyn Middleton took over the directorship from Captain John de Winton, things changed markedly in my favour.

Lyn was a man after my own heart. He was a very direct, experienced and brave naval aviator who stood up for what he believed and didn't suffer fools gladly. As a result, he was both respected and feared by some of his subordinates. Following the CVA-01 decision, he had very little time for the Royal Air Force and, because my job also entailed writing formal DNAW papers for his approval to denounce RAF claims of global tactical air reach, he became my trusted mentor. There was very little that we disagreed about. I wasn't afraid of him, and he liked that. Having read one of my papers that was more than seventy pages long, he asked me to come and discuss the same, which I willingly did.

The paper analysed the claim by the RAF that 'the new Tornado Air Defence Vehicle would be able to provide 24/7 combat air patrol over a Naval Force deployed up to 700 nautical miles from the UK base'. It found that, at best, only one Tornado fighter could maintain a continuous presence over the force at a maximum range of just 400 nautical miles and that effort would require the support of UK's entire fleet of air-to-air refuelling aircraft – leaving the homeland base bereft of any other fighter air defence capability including that of Quick Reaction Alert.

Lyn was a stickler for detail and had only one adverse comment about the paper. I remember the exchange of views going something like this.

'Sharkey, just one thing, on page 64, line 11, I think you have used a wrong word!'

'Really Sir, I beg to disagree. It is the right word within the context of the sentence.'

He grinned, was not at all annoyed and just told me, 'And I beg to differ. Go ahead and change the word. By the way, it is an excellent paper. Well done!'

It was indeed a very good paper but the senior Naval Staff, who tended to distrust their aviators, refused to pass it on to the minister. They didn't want to 'rock the boat' with the RAF. They were more inclined towards appeasement than fighting the naval aviation corner.

Fortunately, and during the 1970s, it had become clear even to our ministers and civil servants that without embarked tactical fighters within a battle group, Soviet Bear reconnaissance aircraft could shadow the Fleet with impunity and relay its position to the squadrons of Badger and Blinder bombers that would round the North Cape if and when hostilities broke out. With grinding teeth, MoD/Air agreed that a limited-capability fighter aircraft should be procured to operate from the Navy's new 'through deck cruisers' that were supposed to be carrying anti-submarine helicopters. The MoD insisted that the role of the new aircraft would be limited to 'hacking the shad' (i.e., deterring and shooting down shadowing Bear aircraft).

And thus, the Sea Harrier Fighter, Reconnaissance, Strike Mk.1 was conceived.

My Sea Harrier intercepting the Bear – with an F-14 in company.

# Chapter 7

# Rising from the Ashes – The Sea Harrier World

By mid-1979, my job as the Sea Harrier desk officer in MoD/DNAW had ended. My predecessor, Nick Harris, had handed over to me a thriving, well-run project and the new aircraft was now ready to enter service for intensive trials and operational development. The leading project defence contractor, British Aerospace, led by John Fozard and Chief Test Pilot John Farley, had worked wonders transforming the ground-attack Harrier into a Fleet air defence fighter aircraft. Greg Stewart and Ferranti had produced the small but remarkable in capability Blue Fox air-to-air, mono-pulse radar, and Andy Cameron of Smiths Industries had produced an outstanding weapon-aiming computer that was seamlessly integrated with a head-up display and an INAS-standard NAVHARS (Navigation and Horizontal Attitude Reference System).

I am proud to record that the £100 million project for the first twelve aircraft (£8 million per unit), controlled and managed entirely by the Naval Staff, was on cost and on time, a significant first in British military aviation history. But was it money well spent? In-service trials and operational development would answer that question. (It is relevant to note here that the RAF Tornado Project for more than 300 land-based fighter and ground attack aircraft had already been approved at a starting cost of approximately £13 billion (£43 million per unit). Time was to tell which of these investments would pay off in terms of strategic and tactical utility/effectiveness.)

Before leaving the Ministry of Defence, I received some very good news. 'Sharkey, you are being appointed as the Commanding Officer of the Sea Harrier Intensive Flying Trials Unit, 700A Squadron.' The words of my naval officer appointer, Robin Shercliffe, were music to my ears. He added, 'We have chosen you over Tim Gedge.'

I have since been informed that Tim did not welcome my appointment with open arms. He had been my predecessor as the senior pilot of 892 Naval Air Squadron. A fine, very conventional naval officer and Qualified Flying Instructor, he has been described as having 'a more moderate approach to operational flying than yours truly' – as stated in a *Spectator* review of the book *Harrier 809*. He certainly did not appear to ruffle any feathers with his superiors and, unlike the Air Warfare Instructors of the Fleet Air Arm, was more attuned to domestic flying practice excellence rather than aggressive war-fighting and pushing the limits. And it was 'pushing the limits' that proved to be vital to the successful conduct of the Falklands air-to-air war.

I suspect that Lyn Middleton would have played a major hand in my new command appointment – for which I am eternally grateful. Had this not been so and had the more moderate Tim Gedge been appointed in my place, I suspect that the Falklands air war may well have been a closer-run thing. A strong statement, yes, but backed up by the events leading to the task force deployment, the events during the war and the post-war testimony by the Carrier Battle Group Commander, Admiral Sir Sandy Woodward: 'If Sharkey Ward had not disobeyed orders we could have or would have lost the war.'

Much to my surprise and delight, I was given a completely free hand by the Ministry of Defence to run the trials unit and subsequent squadrons to my 'own entire satisfaction'. I was ordered to write the defence counsel instruction and my own terms of reference for bringing the Sea Harrier up to a full war-fighting operational standard. This included the production of all standard operating procedures, squadron standing orders, engineering procedures and associated manuals, and last but not least, operational tactical flying training procedures and manuals. I also committed to the submission of a six-weekly report of proceedings and progress to my administrative authority, Flag Officer Naval Air Command (FONAC), which was copied to all other appropriate agencies including MoD/DNAW, A&AEE Boscombe Down and RAE Farnborough.

I'm happy to say the Ministry was as good as its word and let me get on with my task without any interference. My trials team of aircrew and engineers, the majority of whom voluntarily followed me on to 899 Headquarters Squadron and then 801 Squadron, achieved a level of excellence, loyalty and dedication rarely seen.

By the end of 1981, my Sea Harrier squadrons had achieved much. We had the best flight safety record of any fast jet entering service in British military history. We had achieved the reputation, through convincing dominance in the air over much-respected dissimilar NATO fighter squadrons (including the Bitburg, Germany, F-15 and the USAF Aggressor Squadron F-5E), of being the best within-visual-range air combat/dog-fighting aircraft in the Western world – something that the Argentinians had noted carefully. Against all advice from the test pilot world, we had also pioneered and become proficient at night flying from the deck in all weathers. And my engineers and maintenance ratings led by Dick Goodenough had produced an aircraft serviceability rating of second to none. Add to that some very impressive air defence exercises with the United States carrier strike groups – one US Navy admiral signalling my squadron, 'Congratulations to the little jet with the big radar!' – and we knew we had an aircraft with which we could go to war and win.

My pride in my squadron personnel was intense and very well earned. Thanks to them I had become known as Mister Sea Harrier. Not surprisingly, I believed that the standards that we had set would also be achieved by our front-line sister squadron, 800, commanded

No. 801 Squadron Sea Harriers, 1982 – 'the Black Death'.

by Tim Gedge. He had been appointed the first front-line squadron Sea Harrier commanding officer and my team in the Headquarters Squadron had provided operational flying training to him, all his aircrew and all his engineers and maintenance ratings. On the commissioning of his squadron, I had offered Tim full support, especially with the aircraft's state-of-the-art avionics and weapon system. That offer was firmly rebuffed: 'We don't need your help, thank you!' I thought little of it at the time. Inter-squadron rivalry was totally understandable and understood, but as we progressed into 1982, I was informed by my aircrew and by new recruits from the Headquarters Squadron that behind the scenes my integrity and that of my team was being seriously challenged.

Unbeknown to me, 800 Squadron was achieving very poor results with the handling of the aircraft radar and the aircraft's navigation system. My Reports of Proceedings to higher authority had indicated Blue Fox radar contact ranges (23 nautical miles on another Sea Harrier) that were well in excess of Ferranti's own trials and expectations (18 nautical miles) – and contact ranges on larger targets, e.g. Soviet bombers and civilian 747 airliners, varied between 60 and 80 nautical miles. Our navigation computer was also very reliable and produced an accuracy of just 1 mile error every 500 miles flown. Eventually, it was revealed that 800 Squadron could not achieve regular contact ranges with their radar on a Sea Harrier of more than 7 nautical miles and that their navigation system was unreliable and producing worrying results.

I was not aware of this problem until my squadron received new pilots from the shore-based Headquarters Squadron when the Falklands Task Force mobilised early in April 1982. Lieutenant Charlie Cantan announced to all of us gathered in the crew room that 800 Squadron had been spreading the word that, 'Sharkey Ward must be lying in his Reports concerning the capabilities of the aircraft radar and navigation system.' The Command Qualified Flying Instructor, Mike Broadwater, immediately confirmed to me that Charlie's comments were true. This infamous misinformation had apparently become a covert talking point throughout the offices of Flag Officer Naval Air Command and at my home base, RNAS Yeovilton. It had resulted in some heated discussions ashore, but nobody had the decency to address the matter directly to me.

I was not apprised of this unfortunate situation until the moment we set sail from Portsmouth for the South Atlantic. Why do I raise this issue now? I do so because 800 Squadron's distrust of the aircraft's main weapon system was arguably to have a detrimental effect on the conduct of the coming air war.

Tim had relinquished command of his squadron at the beginning of 1982, but his legacy lived on with his relief, Andy Auld, who had the temerity to call me a liar to my face during the journey south regarding the capabilities of the Blue Fox radar. Auld was neither a Qualified Flying Instructor (QFI) nor an Air Warfare Instructor (AWI). He was vociferous about his belief that the fixed-wing Fleet Air Arm did not need AWIs, and that the latter were 'a waste of space'.

A further regrettable factor of which I had not been aware was the disclosure that on sailing from Portsmouth 800 Squadron did not have a single night deck-qualified pilot. They now had eighteen pilots and twelve aircraft in HMS *Hermes*. My 801 Squadron in HMS *Invincible* had just eleven pilots and eight aircraft. Some of my pilots were already fully night deck qualified and all would be available for night flying and night deck alert by the time we reached the Falklands Total Exclusion Zone.

This disparity in operational capability was to have serious consequences when diplomacy failed and hostilities became the order of the day.

My Director from DNAW, Lyn Middleton, was now the Captain of HMS *Hermes* and Flag Captain to Rear Admiral Sandy Woodward. As such, Lyn had considerable influence on Carrier Battle Group proceedings but his only exposure to the front-line operational capability of the Sea Harrier was provided by his embarked squadron, 800 NAS. The squadron did not have any confidence in the autonomous air intercept capability of the Sea Harrier, and their misguided advice and performance when embarked persuaded Middleton to the same point of view. Admiral Sandy and his staff had no other source of reliable advice available in his Flagship and was, in turn, persuaded to the same viewpoint.

As the Carrier Battle Group sailed from Portsmouth on 4 April 1982, it is a regrettable fact that the two carriers, *Hermes* and *Invincible*, were on a different page with regard to understanding the true capability of

801 Squadron maintainers – happy and ready for action.

the Sea Harrier for providing air defence of the task force. This was not helped by Air Vice Marshal 'Blue Rinse' Menaul and other prophets of doom appearing on television on the Saturday before we sailed and stating that 'the Task Force had no fighter air defence and would therefore fail to recover the Falkland Islands' (his nickname came from dying his hair pale blue – and he should've stuck to that). There was no mention in the media of the failed RAF promise to provide air defence to the Fleet throughout the global commons.

Behind the scenes in Whitehall there were those at the highest level who continued to believe in that failed promise. This was evidenced in early 1982 by the public announcement from the Secretary of State for Defence John Nott that our new aircraft carrier, HMS *Invincible*, was to be sold to the Australian Navy. The ignorance of this man concerning the strategic utility and importance of our Fleet was revealed in a remarkable minute recorded in the Churchill College Archives to his eventual successor. The quality of his English and his understanding of UK's global maritime security interests leaves much to be desired, but his intent was clear:

> Defence Minister John Nott to Michael Heseltine (Churchill College Archives.)
> (Sent before April 1982)
>
> I was determined that the Navy was not going to have a greater share of the Defence cake and he [Chief of the Naval Staff, Sir Henry Leach] was equally determined that the plans, which were over ambitious on the Naval side, should be retained.
>
> I had been accused of wishing to ruin the Royal Navy. This charge was totally untrue. Changes that I proposed were best for the Royal Navy. They were just my whim … But I really didn't believe that Naval gunfire support was all that valuable … Now for another decade until the gun seems redundant, although, again, you come back to the central question of strategy, are we to have a Navy which is appropriately designed to fight low-level wars against the Hottentots and a Navy which is to be appropriate if need be for invading the uninhabited islands like Anguilla, or are we to have a Navy which is actually designed to be the main threat of the British people … Which is principally 35–40 miles off the coast of Britain and the nation's flank from Soviet threat.

After the war, he sent another minute to his successor, Michael Heseltine, on defence expenditure:

> I wish you all the best for your reforms but of course all my plans were mucked up by the damn Falklands War.

In the War Cabinet deliberations immediately after Argentina had invaded the Falklands, he was visibly frustrated when the Chiefs of the General Staff and the Air Staff admitted to Margaret Thatcher that they were powerless to intervene with the invasion. And he was even more frustrated and angry when the Chief of the Naval staff, Sir Henry Leach, advised the Prime Minister that the Royal Navy could and would mobilise a task force and recover the Islands. (As disclosed to me by

No. 801 Squadron embarks in HMS *Invincible*, 1981.

Major General Sir Jeremy Moore, War Cabinet member and Land Forces Commander, Falklands.)

On board HMS *Invincible*, our outstanding Captain, J.J. Black, had no delusions about the strategic role of the Fleet and the capability of his embarked Sea Harrier Squadron, 801 NAS.

# Chapter 8

# The Invincible Team Prepares to Engage the Enemy

The first front-line Sea Harrier squadron, 800 NAS, had embarked in HMS *Invincible* under the command of Captain Mike Livesay and Commander 'Air', Neil Rankin more than six months before the commissioning of my squadron, 801 NAS. On commissioning, we were formally designated to be the *Invincible* squadron and 800 NAS was moved to HMS *Hermes*. Although they had received their Blue Fox radars well before 801, we were also formally designated the Blue Fox Radar Trials Squadron. (My team had already overseen the initial radar trials as part of the Sea Harrier Headquarters Squadron, 899.)

Very rapidly, a bond of trust and dedication was created between my Sea Harrier boys and the ship's company, including especially the command. It was a very happy ship. Mike Livesay had been an Air Direction Officer and Neil Rankin was a much-respected fighter pilot and Air Warfare Instructor. Both had vast experience of fixed-wing carrier operations. Neil Rankin had been one of my instructors during my tactical flying training at Royal Naval Air Station Brawdy and also my Commander 'Air' at our home base, Royal Naval Air Station Yeovilton. There was therefore great mutual trust and a willingness to push the boundaries of what we needed and expected to give the Sea Harrier a decent war-fighting capability.

This provided the foundation for demonstrating the ability of front-line aircrew to conduct safe and reliable operations to and from the deck by night and in all weathers, in clear contradiction of the established view from the test pilot world and Controller Aircraft that 'night deck landing in the single seat Sea Harrier was not possible or feasible for the average front-line pilot'. Indeed, the Controller Aircraft had refused to give a formal clearance allowing the Navy to conduct such night deck operations.

With the strong support of my AWI, Ian Mortimer, a long-time Harrier expert, I lobbied the command incessantly to ignore Controller Aircraft's ruling. That pressure bore fruit and the first night deck landing of the front-line Sea Harrier was undertaken by yours truly in June 1981 during an operational inspection of the ship by Flag Officer Flotilla 3. After I landed, Captain Livesay and Neil Rankin brought champagne to the crew room to celebrate this important event. It was thanks to them that the rounded capability of the Sea Harrier as 'an all-weather fighter by day and night' was realised and this contributed significantly to the deterrence of attack and reconnaissance missions flown by the Argentinians against the Carrier Battle Group in the coming air war.

*Invincible*'s unswerving support and encouragement of Sea Harrier operational development was to continue without pause under our extraordinary new Captain J.J. Black and our new Commander 'Air', Dusty Milner – a seasoned fighter pilot in his own right. These four war-fighting naval officers had enabled me to keep my promise to our First Sea Lord, Admiral Sir Henry Leach, when I had the privilege of meeting him at the Pebble Mill at One Studios, Birmingham, having landed the first Sea Harrier in service there on 20 September 1979.

Landing at Pebble Mill at One Studios, Birmingham, 1979.

Sir Henry instructed me as follows: 'Sharkey, the whole of the Royal Navy is watching you. Don't let us down! If there is anything you need just call me and I shall make it so!'

'You have my word, Sir. And thank you.'

I kept my word.

By the time the Carrier Battle Group sailed from Portsmouth, my squadron team was fully prepared and ready to engage the enemy. The pundits in the media had given us no chance of success. Our twenty Sea Harriers would have to face up to 200 Argentinian combat aircraft flown by their navy and their air force. Even our Flag Officer Naval Air Command confided to his staff that he expected 75 per cent of the Sea Harrier aircrew would not be coming home.

On board *Invincible*, the odds against us were foremost in the minds of the ship's company and it took some time to assuage their fears. Those odds were incredibly high, but I was oozing with confidence that, given a fair wind, we could provide sufficient air defence to enable a task force victory. Prior to sailing I gave an interview to the well-known TV news reporter Brian Hanrahan. He listened to what I had to say and then remarked, 'So the Argentinians have much to be afraid of?'

My answer (with an ultra-confident smile): 'Yes, they most certainly do!' I was in 100 per cent war-fighting mode, as was the ship, and that remained in place all the way to the Argentine surrender.

We embarked on the Sunday morning, and, in the evening, my team gave a presentation to J.J. Black outlining the squadron status quo and briefed him on the tactics that we would use to overcome the Argentinian air threat.

Conventional 'iron bomb' attacks against the Battle Group would be deterred and/or engaged with a robust Sea Harrier Combat Air Patrol barrier stationed 100 nautical miles up threat, beyond the surface warship air defence picket line. This barrier would also deter and be effective against the Étendard aircraft carrying the Exocet sea-skimming, anti-ship missile. We had calculated that for the Battle Group to be less vulnerable to Exocet attack, it must remain at least 425 nautical miles from the Argentine mainland. That eased some of J.J.'s understandable concerns. I concluded the brief approximately as follows:

> Sir, there are 5 tasks needing urgent attention. The first, to carry out all the weapons delivery trials that have not yet

> been conducted by A&AEE Boscombe Down. My AWIs have already prepared the trials programme which will last 3 days, and it needs to be carried out as soon as possible. The second, to bring the 4 pilots who have joined us from the Headquarters Squadron up to front-line speed. They are all very experienced and capable aviators and I'm sure they will rapidly adapt to our proven Embarked Operational Procedures. The third, to expand our night team so that all pilots are night qualified by the time we reach the combat zone. The fourth, to hone the air combat and air intercept capability of each pilot to an even higher standard than already achieved. The 5th, to monitor closely the arrival and distribution of the Sidewinder AIM-9L within the Battle Group. I have already prompted FOF3 to arrange and expedite this. May I suggest that we commence the weapons delivery trials as soon as possible on leaving harbour?

J.J. was very pleased with the briefing. 'Well done, 801. You may go ahead as planned. Have a copy of the briefing sent over to the Flagship.' We did so but received no acknowledgment.

Within an hour of leaving Portsmouth, we had delivered a live 1,000lb bomb on a small weapons range to the east. By the time the Battle Group cleared the Bay of Biscay we had carried out all the live weapons trials successfully. These included: lay down, dive and toss bombing; 2-inch rocket projectile delivery; Sidewinder AIM-9G firing against a flare target (we were eagerly awaiting the arrival of the more capable Sidewinder, AIM-9L); tossing a Lepus flare. (Our 30mm Aden cannon had already received full clearance for air-to-ground and air-to-air release.) The aircraft's weapon aiming computer proved very accurate with just one exception. Lepus flares would have to be delivered manually (i.e., the pilot controlling the release range to the target based on Blue Fox radar information).

It was exciting and rewarding for the ship's company to be able to witness all the live firings from the flight deck. The squadron armourers had done an outstanding job preparing all the weapons for flight and delivery with no misfires or hang-ups. The smell of cordite was welcomed by all the sailors, who were becoming more confident every day about the task that lay ahead. Members of the British press were also embarked and were suitably impressed. Our much-loved captain, J.J. gave them half an hour of his very busy time each day to brief them and answer their questions. I was permitted to engage in psychological warfare, giving the press daily quotes about Sea Harriers' capability and how we were going to thrash the living daylights out of the Argentine pilots. These jingoistic quotes were not lost on the Argentinian High Command.

We wrote a detailed report on the weapons trials for the captain and authorities ashore, enumerating the number of weapons released, the required sighting angle for each delivery (provided by my AWIs) and the accuracy achieved – which was excellent in all cases. A copy was sent to the Flagship.

Although again, no acknowledgment was received from *Hermes*, the CO 800 Squadron visited *Invincible* during the long journey south and complained to me about the report, 'Why did you send over that report? You caused me and my team an awful lot of work. Lyn Middleton insisted that we repeat all the trials!' I was dumbfounded. We were going to war, but he didn't like having to prepare for it.

Immediately following the trials, my senior pilot and lifelong friend, Robin Kent, and my AWI, Ian Mortimer, ensured that all the pilots were well briefed, and each was able to experience the delivery of live weapons.

We were well under way towards being 100 per cent ready and we were eager for combat. Night flying qualification for those who needed it was progressing well and our new pilots from the Headquarters Squadron were becoming rapidly attuned to the pace of intensive front-line flying.

The author, Robin and Ian, 1982.

# Chapter 9

# There Emerge Important Conflicts of Opinion Between the Two Carriers

In *Invincible*, J.J. instituted a twice-daily command briefing at 0800 and 1800, to be attended by all heads of departments and including the Sea King ASW Helicopter Squadron CO, Ralph Wykes-Sneyd, and myself. This proved to be of great benefit to the operational and training progress of the ship and its Air Group. J.J. was kept fully informed on both counts and, at the same time, he was able to apprise us all of the stream of instructions emanating from the Flagship – many of which were confusing, frustrating and counterproductive. It rapidly became clear that the Flagship's view of Carrier Battle Group air defence capability and utility differed markedly from our own. This often led to amusing albeit serious discussions in the command briefing about the way ahead for our fixed-wing air group, often ending with the question from J.J., 'CO 801, are you blowing smoke up my arse again?' – delivered with a wicked but entirely supportive smile. My inevitable answer would be, 'No, sir!'

With some hindsight, the differences between the Flagship and *Invincible* resulted from the following.

The Flag Captain, Lyn Middleton, was an extremely experienced Scimitar and Buccaneer fighter-bomber pilot and so his main expertise lay in air-to-surface and air-to-ground operations. This was significant because his only source of knowledge concerning Sea Harrier air-to-air intercept capability came from his embarked squadron, 800 NAS. The latter had openly denigrated the utility and reliability of the Blue Fox radar and the associated weapon-aiming computer, head-up display and navigation system. Quite naturally, this led to the Flag Captain having the same erroneous viewpoint. The total lack of any Sea Harrier night deck landing experience and expertise within his squadron was to play a significant part in the way he allocated associated tasks. (With hindsight, this would account for the disproportionate night deck alert tasking of my squadron during the coming war.)

A key point surrounding this and unfolding events as we proceeded south towards what was to be the declared Total Exclusion Zone (TEZ) is that 'an operational task given by a superior entity, whether the Flag himself, his Flag Captain or the Flag's Staff, amounts to a direct order'. This task/order may be questioned by a subordinate entity, but the superior can tell the subordinate, 'make it so' and that is the end of discussion – one is obliged to do as bid.

This meant that the Invincible Air Group was effectively controlled by the Flagship.

Very rapidly, last-minute signals began to flow from *Hermes* ordering, for example, air defence exercises (ADEX) that entailed the full use of all our fixed-wing Sea Harrier assets. Such a signal would be received without warning and without any consideration for the pre-planned training and aircraft maintenance routines of our squadron. Worst would follow. The ordered ADEX would generally be poorly conceived and would add no operational training value for our air group or surface warships. This was extremely frustrating for *Invincible*, who had no choice but to comply. Our only satisfaction came from completely defeating *Hermes*' Sea Harriers in the air, whether defending our ship or attacking the Flagship.

A combined war planning meeting on board HMS *Hermes* provided us with the opportunity to learn more about the level of training and capability of her squadron – and to put a marker down that was to facilitate *Invincible*'s achievement of self-determination and control during the coming war.

I accompanied J.J. and his ops team to the planning meeting and summarised the level of training and readiness of my squadron. In doing so, I suggested that 801 should conduct and lead an eight-aircraft coordinated strike against Stanley airfield in the small hours of the night at the opening of hostilities – taking the enemy by surprise. I commented that we could boost this number to twelve aircraft if *Hermes* could provide the additional four. I added that we had already successfully practised such an attack against *Invincible*'s towed splash target.

Lyn Middleton sat there in disbelief and then protested, with considerable vigour and contempt, that 'such a coordinated attack by night in single seat aircraft is quite impossible – I should know, with all my twin seat Buccaneer night ground attack experience'. A chorus of agreement from his team backed up their captain, who addressed J.J. and said, 'What do you think, Jeremy?' J.J.'s response was very swift and to the point.

'Sharkey Ward may be as mad as a hatter, Lyn, but if he says he can do something, he can – and you better believe it!'

After much muttering from the *Hermes* crowd, the Flag Captain agreed that 801 should do a demonstration live weapon coordinated attack by night on *Hermes*' towed splash target. But he wasn't happy. (Later, our demonstration attack was in the process of taking place with my Sea Harriers running in at low level from different directions when, at the very last minute, *Hermes* called up and said, 'Trident Leader, you must abort your attack because we haven't got a splash target streamed for you! Sorry about that!' One-upmanship appeared to be the order of the day and was to continue.)

At this stage of the planning meeting, I had a brainwave and made the following suggestion to the Flag Captain. 'Sir, as your Squadron is training hard for surface and ground attack, may I suggest that 800 Squadron takes the lead in that area and that *Invincible* takes the lead in the air defence of the task force. That would of course entail *Invincible* being formally given the authority as the Anti-Air Warfare Control ship, the AAWC.' I wasn't at all surprised when the Flag Captain, with his fighter-bomber background, agreed to the suggestion. The next day, the admiral officially appointed HMS *Invincible* as the AAWC.

After the planning meeting was adjourned, J.J. took me to one side and told me I had taken him by surprise as we had not discussed the AAWC question in any way. I apologised and explained it was a spur of the moment decision that should serve us well and let us get on with the main job. 'Actually, Sharkey, I believe you have done the right thing!'

The appointment of *Invincible* as the AAWC was to have a major beneficial effect on the course of the air war – despite interminable resistance and interference by the Flagship. (This enabled us to mount very low-level combat air patrol missions up threat of the San Carlos Water beachhead. These missions deterred ingress of many of the enemy fighter-bombers, reducing their success rate against our forces.)

As our Carrier Battle Group approached closer to Falkland Islands waters, an Argentinian Boeing 707 intelligence-gathering aircraft began monitoring Carrier Battle Group progress on a daily basis, despite being intercepted and shadowed by Sea Harrier aircraft. As soon as Whitehall gave clearance to shoot the aircraft down, it never appeared again. But it was a sign that the outbreak of hostilities was imminent. Soon we were in those waters, to the east of the Falklands and not far from Tierra del Fuego and Cape Horn.

The Flag had ordered that two Sea Harrier night deck alerts were to be kept ready to go on each deck throughout the dark hours, and two further pilots were to be kept awake and ready to scramble in the crew room. Despite having far fewer aircraft and pilots, and despite protests to the Flag by J.J., 801 was tasked with covering more than half of the night. Fulfilling the task would require the services of at least eight of my pilots. Night deck alert required the pilot to be in the cockpit with all systems ready to go, for a period of ninety minutes – without any heating and with an outside air temperature averaging 2°C.

We were already flying continuous combat air patrol missions throughout daylight hours and this new tasking took no account of aircraft

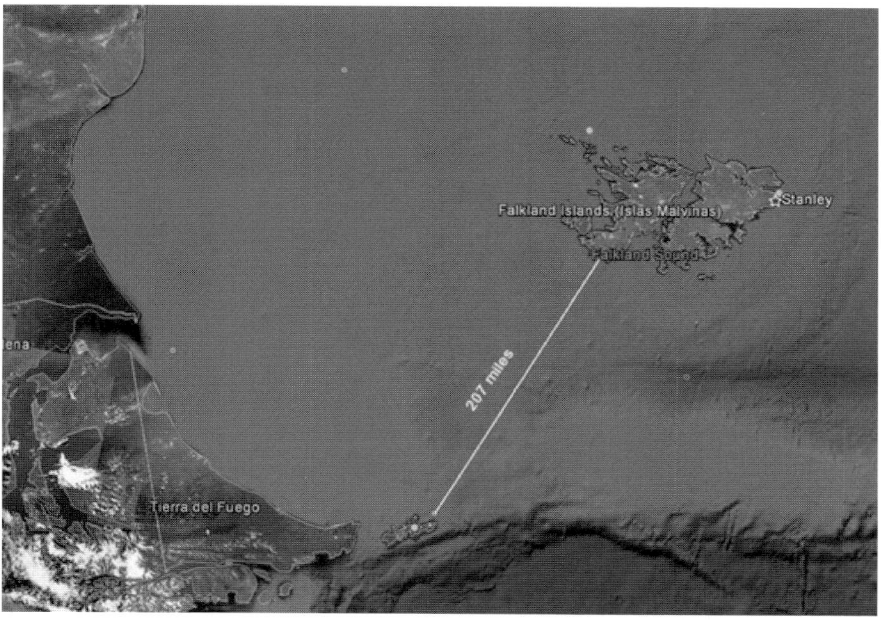


Eight thousand miles from home – on a wing and a prayer.

and pilot strengths, or the inevitable, serious pilot fatigue that it would entail. At the command briefing in *Invincible*, I addressed this matter to J.J. and Dusty Milner. 'My squadron assets are too small to fulfil this tasking, Sir.' I explained the problem in detail – my pilots would get no rest at all. 'I propose that we ignore the Flag's instruction and do it my way.' This was to be one of the many times that I successfully suggested disobeying orders. Dusty wasn't too keen on it, but J.J. listened carefully, saying, 'and what is that, Sharkey?'

'One alert aircraft on deck and a second pilot in a sleeping bag in the crew room!'

J.J. could see the common sense of it all. 'Very good, Sharkey. Make it so! We shall do it your way!'

Before April was over, the rules of engagement (ROE) for the task force were signalled from London. J.J. called me to the bridge. He was not amused. 'Have a look at these, Sharkey and let me know what you think. I can't make head or tail of them!' In brief, the ROE stated that we were not allowed to open fire on the enemy unless he had fired first. A completely insane instruction, especially regarding air-to-air combat! When I went through the contents of the ROE with all my pilots, they were aghast. 'Don't worry!' I told them, 'From now on you will obey my own ROE and I shall take full responsibility. If anything moves in the sky that is clearly not friendly, i.e., enemy aircraft, you are to immediately engage it and shoot it down. Is that clear?' My pilots cheered and I went up to see the captain. 'All done and dusted, Sir! I have torn up the signal and told my team that they are to do it my way.' J.J.'s eyes twinkled.

I was starting to believe we were up against more than just one enemy. This was reinforced on the eve of 1 May at the command briefing. The squadron was all geared up for conducting the night attack on Port Stanley airfield that we had trained so hard for.

J.J.: 'I'm afraid I have bad news, CO 801. Your night attack is not to take place. Instead, Hermes aircraft are to strike the airfield and the airstrip at Goose Green at dawn. Further, a loan Vulcan will conduct a high-level bombing raid on the airfield in the small hours of the night. You are to provide it with airborne fighter cover. No doubt you will fly that mission yourself.'

J.J. could see the disappointment and frustration in my face. 'But there is more bad news. 801 is to carry out all the night alerts on deck so that Hermes aircrew can rest. And to cap it all, our Battle Group surface warships are not allowed to fire at anything in the air while the Vulcan is in the area.'

There was little I could say. I was angry and disappointed. 'Roger, Sir! Do I take it we shall conduct the alert on deck my way?'

'Certainly, Sharkey!'

'Aye, aye, Sir.'

I retired to my cabin thinking, what the hell is this? A dawn strike? How original? Are we taking part in a John Wayne movie?

The morrow was going to be a significant day in the Falklands Air War, and I needed to get some sleep.

# Chapter 10

# 1 May 1982 – A Pivotal Day in the Air War

Thorough preparation is vital to the successful conduct of any event or task. *Invincible* had indulged in such preparation and was now ready to engage the enemy more closely.

The Vulcan bombing raid came and went as I held airborne vigil under the stars, hoping to get to grips with enemy fighters. Non appeared. The first of the Vulcan bombs clipped the side of the Port Stanley airfield runway but failed to prevent air operations continuing. The remaining twenty bombs missed the target entirely.

That was no surprise because the RAF planning team ashore had calculated that more than twenty such missions would be needed to successfully interdict the runway. Even then, bomb craters could be filled in rapidly.

Immediately following the unsuccessful albeit long-in-duration bombing raid, several of our frigates and destroyers conducted intense naval gunfire bombardment of the airfield. As dawn broke and with the airfield defensive positions fully awake and ready, the twelve 800 Squadron Sea Harriers from *Hermes* delivered courageous low-level attacks against the airfield and the airstrip at Goose Green. Surface-to-air gunfire inflicted battle damage on some of the aircraft but, thankfully, they all returned safely to the deck.

801 conducted continuous combat air patrol (CAP) missions from before dawn until after dark and our well-prepared aircraft weapon systems and pilot tactics lived up to our most optimistic expectations – thoroughly discounting the misinformation that had been spread by our sister squadron.

Waves of high-level, high-speed Mirage III fighters under the control of land-based air-direction radars approached our medium-level CAP positions (each of which consisted of two Sea Harrier aircraft) and

Vulcan bombs fail to interdict the runway at Port Stanley.

conducted rather tentative slashing attacks towards each of the CAP pairs. As soon as our Sea Harriers turned and climbed to engage them, they turned away and ran for home at full speed. But not before they had jettisoned their external fuel tanks, which left a clearly visible white trail of unspent fuel behind them. It was easy, in the heat of the action, to confuse these trails with the exhaust from missiles. There were few if any replacement fuel tanks available for the Mirages. This led to them having insufficient range to progress past Falkland Sound towards the Carrier Battle Group – a self-inflicted wound that was to limit their operational utility. (They were not moved to the north to protect Buenos Aires from 'possible' Vulcan attack. That was wishful thinking by the RAF.)

Low-level Mirage V Dagger fighter-bombers, unseen by the Battle Group's air defence radars, attacked our warships close to Port Stanley and some of these were brought down by the Sea Wolf short-range missile system. Almost simultaneously, a key event took place in the upper air.

A pair of Mirage III fighters decided to pursue their slashing, supersonic attack on the 801 CAP pair of Steve Thomas and Paul Barton. The two Sea Harriers climbed to meet them, and Thomas gained radar contact head on at 13 miles. This was an excellent detection because the Mirage radar echoing area is very small.

Things happened very fast. 801's well-practised fighter combat tactics came into play. Barton immediately took lateral separation from Thomas, preparing to intercept the Mirages from their beam while Thomas tried to get head-on acquisition with his Sidewinder AIM-9L missile. But the Mirages must have throttled back to hide their hot exhaust – the missile would not acquire the target. As he flew between them and turned hard to engage, the Mirages jettisoned their fuel tanks and took evasive action, one going high and the other one spiralling down towards the clouds. In the classic fashion of this 'hook' procedure (known in US Navy parlance as 'the Polish heart manoeuvre'), Barton closed from the beam and released his Sidewinder at the higher Mirage. Direct hit, a ball of flame and the first air-to-air kill of the war. Seconds later, Thomas released his Sidewinder in a near vertical dive and claimed the second kill of the war. All the preparation and combat training had paid off handsomely. Excellent work by both pilots!

Very shortly afterwards, a Mirage V Dagger pulled up directly in front of Bertie Penfold of 800 Squadron, after attacking our ships near Port Stanley. His Sidewinder AIM-9L claimed the third kill of the war. It was jubilation all round!

The icing on the cake was yet to follow. Just as dusk was falling, Alan Curtis and Mike Broadwater of 801 carried out a fully autonomous intercept of three Canberra bombers approaching the Battle Group from the north-east at very low level. Curtis' Blue Fox radar gained very good contact at more than 20 miles looking down over the sea. He took some lateral separation and then guided his CAP pair into a firing position on the beam of the Canberras. They were in and out of cloud. His Sidewinder destroyed the nearest Canberra, but the other two bombers managed to escape in the cloud. A perfect intercept with the Sea Harrier full weapon system.

The two successful 801 Squadron engagements were of pivotal importance. It has been reported that the commanding officer of the Mirage III squadron returned to base and immediately proclaimed to his High Command that 'his team could not win against the Sea Harrier and the AIM-9L Sidewinder missile'. In the heat of the moment, he was castigated for his viewpoint, but the Argentinian High Command had taken note of what they already feared. They knew that the Sea Harrier was outstanding in a dogfight and the AIM-9L was seen by them as an ace card. They instructed all their fighter and fighter bomber pilots to avoid the Sea Harrier at all costs, even if it meant aborting attack missions, ditching their weapons and returning to base to try again another day. This occurred on many occasions after the amphibious force had established the beachhead in San Carlos Water.

The Argentinian media christened the Sea Harrier *La Muerte Negra* (the Black Death). This was not lost on the *Invincible* command team, now the Anti-Air Warfare Control Ship (AAWC). 'Fear of the Sea Harrier' was to prove an effective deterrent to the air attacks on our Carrier Battle Group and the San Carlos Water beachhead. That depended upon keeping our CAP stations manned and ready in the direction of the threat. But that was not always possible despite explicit instructions from the AAWC to both carrier air groups.

At the command briefing that evening, J.J. applauded the pilots who had done so well, then added, 'but, sadly the Flagship has not even commented on our kills. Further, Sharkey, we have been tasked once more with more than half tonight's night deck alert, two aircraft manned on deck and two pilots ready to scramble in the crew room. Any comment before I go down to the wardroom to congratulate your pilots?'

'Yes, sir. Not unexpected. Shall we do it my way?'

'Definitely. Make it so!'

# Chapter 11

# The First Week of Fighting Sets the Tone

Before closing the command briefing, J.J. informed us that the Argentine Exocet-armed cruiser ARA *General Belgrano* had sailed with its Surface Battle Group. It was loitering to the south-west of our Carrier Battle Group and was being shadowed by one of our nuclear submarines, HMS *Conqueror*. It was also understood that the Argentine aircraft carrier, ARA *Veinticinco de Mayo* had put to sea and was probably somewhere to the north-west – indicating a possible pincer movement against us.

'I'd like to send one of your aircraft on a probe mission to check the whereabouts of their carrier, Sharkey. Can do?'

'Of course, Sir. I shall get Ian Mortimer to fly the mission. I would normally do it myself, but I'm absolutely shattered.' Fatigue was already taking its toll.

And so it was, as I lay sleeping, Mortimer flew an extremely dangerous hi-lo-high mission out to more than 200 nautical miles from *Invincible* and found the *Veinticinco de Mayo* and her escorts. He was at very low level as he used his radar to accurately mark the position of the enemy Carrier Battle Group on his navigation computer, was illuminated by several fire control systems but escaped and returned to the ship. A truly brilliant mission execution! The enemy position was immediately signalled to the Flagship but, amazingly, no response or follow-up action took place.

However, on 2 May, the War Cabinet in UK gave the order to the nuclear hunter killer submarine, HMS *Conqueror*, 'Sink the Belgrano!' The loss of the Argentine cruiser resulted in the whole of the Argentinian surface fleet returning to harbour, including the *Veinticinco de Mayo*: from where they never ventured until after the end of the war. But the air warfare initiative against our Battle Group was to continue with some notable success by the enemy. This was not to be at the hands of direct attacks by Mirages, Daggers and Skyhawks.

Following their losses on 1 May, they didn't resume their major offensive against the Carrier Battle Group and other task force shipping until 21 May, when our Amphibious Group set up the beachhead in San Carlos Water. But the Argentines had what they hoped to be an ace card available, that was deployed with the main intent of sinking one of our carriers. That was the Étendard carrying the sea-skimming, anti-ship Exocet missile.

In the light of this threat, Rear Admiral Sandy Woodward moved his Battle Group to the east, outside the anticipated attack range of the Étendard. But in the next few days, the fog of war descended upon the task force, encouraged by two things.

First, the failure of HMS *Hermes* to keep the Anti-Air Warfare Control ship *Invincible* informed about the air movements of its Sea Harriers, many of which were deployed on nugatory visual searches of the sea areas around the Battle Group. This led to intense confusion for both Sea Harrier CAP aircraft and surface warships.

Second, the ill-conceived second attack against Port Stanley airfield and Goose Green airstrip.

On 4 May, the second Vulcan bombing mission took place in the dark hours of the night, this time missing its target entirely. Naval gunfire bombardment of the airfield followed but the planned dawn strike of twelve *Hermes* Sea Harriers had to be delayed due to very low cloud over the area.

To facilitate the Sea Harrier attack, the admiral moved the Battle Group closer to the islands, leaving the outer screen of air defence warship pickets within Étendard and Exocet range. When the delayed dawn strike took place, Argentine surface-to-air weapon systems were ready to receive it and Lieutenant Nick Taylor was shot down and killed over Goose Green.

Illogical directives continued to flow from the Flag.

The *Invincible* CAP pair of Robin Kent and Brian Haigh were in the process of investigating a fast-moving intermittent contact to the west of our air defence destroyers. The Exocet threat was at its height. Without any discussion with the AAWC, *Hermes* ordered the pair to return to their CAP station having assessed, quite incredibly, from over 100 miles away the fast-moving contact to be 'spurious' – a costly mistake.

This misguided order from *Hermes* was soon to be followed by another one, this time with tragic results.

*Invincible* had stationed three CAP pairs up threat of our air defence destroyer pickets. Again, without any by your leave to the AAWC, *Hermes* ordered the southerly CAP pair led by John Eyton-Jones to leave their station protecting HMS *Sheffield* and proceed 120 nautical miles to the south-west to conduct a visual search for surface warships. Our Blue Fox radar could have sanitised the area in question with just one sweep. Another firm indicator that *Hermes* refused to acknowledge the proven capability and utility of the radar.

Shortly after the CAP had been removed from station, an Étendard closed into the empty station, launched its Exocet and struck HMS *Sheffield* amidships, totally disabling her.

*Hermes*' regrettable distrust of the Blue Fox radar was to have more tragic consequences after the amphibious group had landed.

A further tragedy was to beset our Sea Harrier force on 6 May. When investigating a reported enemy contact at very low level and in thick cloud, two of *Invincible*'s aircraft were lost without trace, presumably resulting from a mid-air collision. Alan Curtis and John Eyton-Jones did not return to the ship. This left us with six aircraft and just nine pilots.

*Hermes*, now with eleven aircraft, displayed no recognition of this, ordering disproportionate levels of night deck alert for my squadron – which we had to ignore.

HMS *Sheffield* immediately after the Exocet attack.

Soon, two more extraordinary signals from the Flag arrived in *Invincible* in quick succession. Both concerned the Blue Fox radar and neither recognised the major successes achieved by the radar on 1 May – without which Argentine fighter aircraft would not have been ordered to avoid contact with Sea Harriers at all costs.

'The Blue Fox Radar is not to be used to look down over the sea for enemy targets.'

And then, 'The Blue Fox Radar is not to be used by aircraft on combat air patrol until a ship-based Air Direction Controller has guided them to within 10 miles of their target.'

The tone for the air war had been set by the admiral's staff – and there was no ambiguity in its intent. No. 800 Squadron's confirmed inability to utilise the Sea Harrier aircraft's weapon system effectively was now prejudicing the successful conduct of our air-to-air war.

I persuaded J.J. Black that these two signals, and others to come, needed to be ignored – and we should conduct the air war my way. He agreed.

# Chapter 12

# The Lull Before the Storm

Between 6 and 21 May, the threat of further Exocet attacks was not taken lightly. Continuous combat air patrols were conducted throughout daylight hours and night deck alert aircraft were kept ready to scramble throughout the dark hours, even though these were adjudged a relatively safe period. In *Invincible*, the shortage of aircraft and pilots had induced serious pilot fatigue. We all awaited the arrival of the Amphibious Task Group commanded by Commodore Mike Clapp with keen anticipation and with it eight reinforcement aircrew and aircraft to be delivered by the ill-fated SS *Atlantic Conveyor*. This container ship was taken up from trade under the civilian command of its Merchant Navy Master Ian North, and under the military command of Captain Mike Layard (later to become Second Sea Lord). He was ably supported by Tim Gedge, 809 Reinforcement Squadron Sea Harrier Commanding Officer. Between them they organised the rapid reconfiguration of the upper deck into a Sea Harrier landing pad and aircraft park.

When the *Atlantic Conveyor* rendezvoused with the Carrier Battle Group on 18 May, *Hermes* and *Invincible* each received on board four Sea Harriers and four pilots.

This signalled immense relief within 801 Squadron. We now had ten aircraft and thirteen pilots and could look forward to providing a more substantive air defence posture for our ground forces led by Brigadier General Julian Thompson, Royal Marines.

Very sadly, the *Atlantic Conveyor* was fatally struck by an Exocet missile soon after arriving on station and before she could disembark her stores, supplies and helicopters at the beachhead. The Étendard pilot mistook her for one of our carriers.

801 Squadron in *Invincible*, with the latter appointed as the AAWC, had already planned the air defence of the beachhead considering the

total number of Sea Harriers available, twenty-five. The Carrier Battle Group would have to keep a safe distance from the Étendard base on the mainland and this would mean long transits from the carriers to the three designated CAP stations to the west, north-west and south-west of San Carlos Water. This meant that time on station would be limited.

Air attacks on the beachhead would be delivered at a time of the enemy's choosing, whereas we would have to mount a continuous presence in defence of the beachhead throughout daylight hours. It was not a completely one-sided equation though. We now knew that Argentine fighter-bomber aircraft had been told to avoid engaging with Sea Harrier at all costs and to abort their mission if we got in the way. This was immensely important and represented a significant deterrent value. If we could sustain the CAP barrier, we could and would turn away some of the attacking aircraft before they could deliver their weapons.

Our *raison d'être* was now to protect our ground forces and I explained to the captain and to my pilots that I was determined to push all the limits to achieve that end.

'After transiting at high level, you are to descend and maintain your combat air patrol at very low level with your radars blasting out continuously to tell the enemy that you are there. Very low level, because this will enable you to detect enemy aircraft at the same height visually. We all know that our radar is unable to detect low-level targets over the land! You will also fly your CAP at endurance speed to extend your time on station. If you get into combat, you are good enough to take on the enemy at any speed. Finally, I know that the distance back to the carrier will probably vary between 125 nautical miles and up to 175. Predicting fuel consumption in our little jet is very easy and accurate and so you are not to leave your CAP Station until you have calculated a land-on fuel state of no more than 800lb. If you take liberties with these instructions, I shall have your guts for garters. Is that clear?'

My aircrew knew I wasn't joking and, all credit to them, there were no complaints.

In our sister carrier and unbeknown to us, the Flag Captain ignored the AAWC's defensive combat air patrol plan and issued his own instructions to his Sea Harrier pilots.

'You are not to fill the CAP stations designated by *Invincible*. Instead, you are to hold station directly above the beachhead at altitude, without use of your radar, and are to be prepared to descend rapidly and engage enemy aircraft when so directed by the on-site Air Direction Controller. Further, you are to return to the deck and land on with no less than 2,200 pounds of fuel.'

As a direct result, throughout the remainder of the air-to-air war the low-level CAP stations that were supposed to be filled by 800 Squadron aircraft were left empty, significantly reducing the CAP barrier's effectiveness as a firm deterrent to enemy air attack. Despite this, 801-manned stations turned away up to 450 fighter-bomber attack missions without them delivering their weapons. This statistic was provided by the Argentinian Embassy in London after the war.

# Chapter 13

# Air Attacks on the Beachhead are Intense – Full Deterrence and Airspace Denial Not Achieved

The Flag's instructions to *Hermes*' Sea Harrier pilots had to be obeyed. Nor were they contested, as they should have been, by 800 Squadron's commanding officer. Strict orders to return to the deck with more than 2,000lb of fuel after their long transits to and from the beachhead, left the pilots with little more than ten minutes time on task, even at high level. (With a target land-on fuel of 800lb, *Invincible*'s low-level CAPs were able to remain on station for between twenty-five and forty-five minutes, depending on the distance from the Carrier Battle Group.)

There was no way that high-level CAPs over the beachhead could visually detect ingressing, low-level, fighter-bomber attacks, and so they had to wait until they were called into action by the Air Direction Officers in the warships of the amphibious force. This meant that they could not engage the enemy until after the enemy had delivered its weapons – 'after the horse had bolted'.

Wave after wave of Mirage Daggers and Royal Navy Skyhawks attacked the beachhead continuously for the next three days – after which time the intensity of attacks slowly diminished following heavy attrition of Argentina's fighter-bomber aircraft.

Commodore Clapp set up a 'ring of steel' outside San Carlos Water in Falkland Sound with seven destroyers and frigates absorbing some of the punishment destined for our landing forces. The attackers passed through empty CAP stations left by *Hermes*' aircraft.

It is telling that only one group of attacking aircraft attempted to pass through one of the three manned low-level CAP stations and that was on 21 May. Three Daggers at very low level were visually sighted

Painting by Robert Taylor of my Dagger kill.

approaching over West Falkland by Steve Thomas and me. We engaged them in combat and shot the three of them down.

No more attempts were made to pass through manned stations. The deterrence plan was working but was severely hindered by the lack of support from the Flagship.

It must be said that the pilots of 800 Squadron acquitted themselves very well whenever they had the opportunity to chase and shoot down retreating enemy attack aircraft.

In the same breath, it must be said that leaving low-level CAP stations vacant allowed more Argentine attack aircraft to reach and deliver their attacks against the targets in San Carlos and surrounding waters. This led directly to the demise of four gallant warships, HMS *Ardent*, HMS *Antelope*, HMS *Coventry* and, of course earlier, HMS *Sheffield*. These casualties could and should have been avoided. It must be noted that our warships defending San Carlos Water were in a land-locked environment for which their surface-to-air weapon systems were not designed. With that in mind, they achieved miracles. The air-to-air war was won dramatically by the Sea Harriers of the Fleet Air Arm with twenty-five kills and non against, yes. But the surface-to-air war was won, equally dramatically, by our surface warships with the same number of kills.

The air-to-air and surface-to-air battle in defence of the amphibious beachhead enabled the safe disembarkation of our Royal Marines-led ground forces without major casualties. Our land forces were heavily outnumbered, but they prevailed and presented the UK with a memorable victory. This was the result of great leadership, outstanding teamwork and professional preparation and training.

On 14 June 1982, Major General Menendez formally surrendered to Major General Jeremy Moore in Port Stanley. Against all the world's predictions to the contrary, the task force had answered the call and won what has been described as 'the last of the conventional wars'. The Carrier Battle Group had clearly demonstrated the need for a balanced Fleet including robust fighter air defence. Admiral Sir Henry Leach summed it up concisely: 'Without the Sea Harrier there could have been no Falklands War.'

I was airborne en route to CAP when I was informed, 'Sharkey, the war is over. Weapons tight!' I didn't believe it at first. Then the news hit

801 Squadron enjoying victory.

Admiral Sir Sandy Woodward GBE KCB.

me squarely between the eyes. Strangely, it felt as though the bottom had dropped out of my world – a world 8,000 miles from home that had been full of frustration, fatigue, dedication and endeavour.

Much later, I became close friends with Admiral Sir Sandy Woodward GBE KCB and worked with him for two decades to try to right the sinking ship that was the Royal Navy. (See my second book, *Her Majesty's Top Gun and the Decline of the Royal Navy*.) He became an avid supporter of the fixed-wing Fleet Air Arm and of the vital part to be played by aircraft carriers in pursuit of UK's global national security interests.

# Chapter 14

# The Strategic Lesson from the Falklands Air War

At the beginning of the 1980s, Argentine dictator Leopoldo Galtieri had taken note of the withdrawal from service of UK's strike carriers and believed that, without them, Her Majesty's Government would be powerless to prevent the occupation and annexation of the sovereign British territory, the Falkland Islands. Then, the public announcement by the UK government's ill-informed and misguided Minister of Defence John Nott, in late 1981/early 1982 that the government was planning to sell our first Sea Harrier carrier, HMS *Invincible*, to Australia, encouraged Galtieri's decision to invade the islands.

He was also fully aware that the RAF's promise 'to provide air defence to the Fleet throughout the oceans of the world with their land-based tactical fighter aircraft' was groundless propaganda. He clearly had more common sense than our Whitehall mandarins and our Ministry of Defence.

The fortunate thing for Britain was that Galtieri got his timing wrong. A few months later in the Antarctic winter, or a year later when the UK would have further diminished the size and capability of the Royal Navy fixed-wing Fleet Air Arm and of our amphibious forces, the end result would have been rather different – perhaps no Operation Corporate but conceding the loss of one of our sovereign territories to an invader.

The overriding strategic lesson to be learned from the Falklands air war is that:

> As an island nation with vital global trade and energy reliance and the desire to protect its sovereign overseas territories, the UK must invest in a robust maritime power projection capability to safeguard those interests. Throughout history, both ancient and modern, the key to that capability has

> been the maintenance of a balanced naval fleet. Since the introduction of manned aircraft to the world, that balance of defensive as well as offensive capability has included the utility, versatility, flexibility and mobility of aircraft carriers – and in the modern sense, strike carriers. That capability cannot be provided by land-based tactical fighter aircraft.

This lesson is reinforced when comparing the strategic utility, cost and operational effectiveness of the Royal Navy Fleet Air Arm Sea Harrier (initial programme cost £100 million) and the land-based Tornado (initial programme cost £13 billion). The former has answered the call, in exemplary fashion. Despite intentionally misguided promises from the RAF, the latter's global expeditionary force utility and operational effectiveness has proved to be non-existent.

There is no suggestion or intent that the UK should invest in a navy that can independently wage global war. But the UK has an obvious vested and moral obligation to be able to contribute to the policing of the global commons and to global or limited warfare in conjunction with our allies, particularly the United States and NATO. We can and should embrace that contribution with a core navy that is powerful, balanced and professional. That is within the means of UK's national resources and would allow us to act independently, when necessary, in the protection of our special global national interests.

In the light of our declared strategic maritime policy, such investment should take precedent over the procurement of other weapon systems that do not have global reach, e.g. large and expensive fleets of land-based tactical aircraft and of heavy tanks and artillery.

# Chapter 15

# Principal Tactical Lessons from the Falklands Air War

Lesson one: maritime expeditionary task forces and/or Carrier Battle groups need to maintain a robust outer layer of air defence up threat of all surface and ground force units. This is as relevant today as it was in 1982, when only a small number of Sea Harrier fighter aircraft were available for this task. They achieved it by filling designated combat air patrol (CAP) stations up threat of the deployed surface warship air defence pickets. When these CAP stations were left unmanned for whatever reason, that is when enemy fighter bomber aircraft and missile carriers conducted their attacks on the Carrier Battle Group and later, on the amphibious beachhead.

The RAF had promised ministers that it could provide air defence of the Fleet throughout the global commons, encouraging the government to invest in large fleets of land-based tactical fighter aircraft. The Falklands air war demonstrated that it was a false promise, but this had already led to insufficient investment in Royal Navy Fleet Air Arm carrier-borne fighter aircraft. The billions of pounds spent on land-based tactical fighter assets proved to be a waste of the taxpayers' money and diverted scarce resources away from the maritime strategic need. This resulted in the Falklands War being 'a close-run thing' with many unnecessary ship and personnel casualties.

The Ministry of Defence failed to take note of the dishonest claims of the RAF and continued to invest hundreds of billions of pounds in the land-based Tornado and Typhoon fighter projects – neither of which produced aircraft capable of global air deterrence and defence over the sea or of strategic maritime power projection. The lesson here is that Whitehall mandarins and politicians, especially defence ministers, should study recent military history more closely and cease listening to the loudest voice of a single service that cannot be trusted.

Lesson two: the operational availability rate of the early Sea Harrier for combat with its relatively simple and easy to maintain weapon system was outstanding (99 per cent in 801 Squadron). This enabled it to provide significant airborne deterrence against the enemy air threat. Such an extraordinary rate is not available to today's more sophisticated state-of-the-art fighter aircraft, including the F-35B (operational availability rate approximately 65 per cent – although this may improve). This new aircraft's exceptional look-down, shoot-down, beyond visual range capability will not be fully realised when entering a combat theatre, unless there are sufficient aircraft embarked to maintain a 24/7 outer layer of combat air patrol capability up threat of expeditionary force units. Should CAP stations be left empty because of the shortage of embarked fighter aircraft, tragic and avoidable losses of surface warships will be the result. The reluctance of the Ministry of Defence to invest in permanently embarked fighter air groups of the necessary size (a peacetime complement of two twelve aircraft squadrons dedicated to each aircraft carrier, building to three squadrons in times of high tension) will prevent the necessary 24/7 capability – and represents a latent threat to our national security interests.

Lesson three: entering into combat, it is a golden rule that one should know one's enemy and his capabilities. Complementing this rule, it is equally important that personnel at all levels of a combatant force should have a complete knowledge and understanding of the capabilities of each of its weapon systems: its strengths and its weaknesses. With such knowledge in hand, one can engage in successful combat by taking advantage of the enemy's weaknesses and capitalising on your own strengths. When 801 Squadron went to war in the Sea Harrier, the aircrew did have such knowledge in hand. They 'knew their aircraft and its weapon systems backwards'. Sadly, when the task force sailed, the staff provided for Rear Admiral Sandy Woodward, Commander of the Carrier Battle Group, had practically zero knowledge of the true capabilities of the Sea Harrier. This was true of the admiral himself, who openly admitted the same. (MoD/Royal Navy had failed to disseminate the formal written reports provided by the Sea Harrier Trials Unit and Headquarters Squadron. These detailed reports provided a full overview of the Sea Harriers' air defence, air intercept and air-to-air combat capability that had been proven in trials.) He and his staff were therefore limited to learning about the full capability of the Sea Harrier through the eyes of the embarked squadron in the Flagship, 800 Squadron.

As related earlier in some detail, the 800 Squadron aircrew did not trust their air-to-air weapon system and, as a result, were unable to use it to best effect. The Flag therefore assumed this to be the norm within the Sea Harrier community and ran the fixed-wing air war accordingly. This in turn left 801 Squadron in HMS *Invincible* in an invidious position: follow the Flag's often ludicrous orders and possibly lose the war or disobey the Flag's orders and win the fight. It was the latter course that 801 Squadron and HMS *Invincible* adopted, despite perpetual interference from the Flag. At the end of the day, the admiral admitted that Operation Corporate was a 'close-run thing' and, at a later date, he expanded on this by stating that, 'If Sharkey Ward had not disobeyed orders we could have/would have lost the war.' The clear lesson here is to know your weapon systems and how best to use them.

Lesson four: during the war, the limitations of:

a. the Sea Harrier weapon system (lacking a long-range, look-down and shoot-down capability, particularly over land) and,
b. our surface warships weapon systems (inhibited/severely degraded in a landlocked environment)

demonstrated a serious lack of investment in the UK's maritime operational capability. This is particularly important in the light of probable expeditionary force operations throughout the global littoral in defence of energy and trade supply routes. The procurement and embarkation of the F-35B aircraft removes the first of the above limitations, but urgent attention is needed for the fitting of state-of-the-art weapon systems to our surface fleet units.

# Chapter 16

# The Most Extraordinary Volte-Face of the Twentieth Century

When the victorious task force combatants returned to the UK, they were given a rapturous welcome by the public. But not all were welcomed so warmly by their military counterparts and superiors.

When Rear Admiral Sandy Woodward, the Commander of the triumphant Carrier Battle Group, returned by air and landed at RAF Brize Norton, he was met by the Commander of the Task Force, Admiral Sir John Fieldhouse, a submariner by trade, who welcomed him warmly

HMS *Invincible* is welcomed back to Portsmouth – but I had already been flown back to the UK to attend Staff College and, sadly, missed the grand occasion.

and then gave him the extraordinary instruction, 'By the way, I want the Falklands Air War victory to be kept very low profile!' Sandy was thoroughly taken aback. The motive behind this instruction remains a complete mystery. Why minimise the vital part played by the Fleet Air Arm fixed-wing Sea Harrier squadrons at a time when the Air Staff in the Ministry of Defence continue to challenge the existence of the same? Did internal partisan politics play a role?

When personnel of the guards regiments landed at Brize Norton, they were met by a line of four-star generals. But when Commodore Michael Clapp, the commander of the victorious Amphibious Task Group, landed, he was shown great discourtesy with not one senior naval officer there to welcome him home.

At a much lower level, when Commander 'Sharkey' Ward landed at Brize Norton in full uniform with a planeload of Welsh Guardsmen and Falklands islanders, he was initially refused entry to the VIP reception lounge (which was already full of all the other passengers) by a self-righteous flight sergeant, who told him, 'You can't come in here, Sir! This is only for Falklands War Veterans!'

A similar welcome awaited him at his home Naval Air Station, Yeovilton. The captain there, one Peter Williams – a helicopter pilot, asked to see him in his office – but there was no warm welcome or congratulations. Instead, he was kept standing and was berated in a very antagonistic manner with the words, 'Listen here. What we don't need on my air station is some cowboy fighter pilot telling us all how his squadron won the air war.' Sharkey turned and left without any reply and wondered, what the hell is going on? Could it be that 800 Squadron, which had returned a month before, had, as they did before the Falklands, spread derogatory rumours about their sister squadron? With hindsight, that would appear to be a possibility.

The sum of it all was that the Silent Service wished to remain silent, at least about the air-to-air war (so as not to annoy the RAF, who played no part in it). Add to that, palpable jealousy emanating from some of those who did not take part down south or who did not perform well, and the stage was well set for the RAF to successfully spread their propaganda that it was they who had 'won the air war in the Falklands'. This disgraceful propaganda initiative continues to this day and has not been challenged adequately by the majority of successive Naval Staffs.

A Falklands memorial service was held in St Paul's Cathedral. Arrangements and invitations were controlled by a wing commander of the RAF Regiment. The cathedral was packed with RAF officers and airmen even though there was just one RAF fatality during the war (a fine young helicopter pilot based in HMS *Hermes*). Some 256 naval and ground force personnel had given their lives, but the invited congregation bore no resemblance to their sacrifice. Tellingly, the commanding officers and personnel of the two victorious Sea Harrier squadrons were not invited.

Later, a banquet was held in the Guild Hall for those decorated for their service during Operation Corporate. Once more, it was arranged and controlled by the RAF and three significant participants in the Falklands War were not invited, i.e., Rear Admiral Sir Sandy Woodward, Carrier Battle Group Commander; Brigadier-General Julian Thompson, Commander 3 Commando Brigade, Royal Marines; and Commander 'Sharkey' Ward. Was it a coincidence that Margaret Thatcher had named these three officers amongst the six most influential 'names' of the Falklands conflict? Perhaps so!

A new and fallacious history of the Falklands air war that did not reflect the truth concerning Operation Corporate was being fabricated by the powerful RAF public relations team. For example, the Junior Service celebrated 'twenty-three surface-to-air kills by their RAF Regiment Rapier squadron'. De facto, they achieved zero kills.

Astonishingly, the Naval Staff appeared to go along with the RAF initiative. The Navy Presentation Team toured the UK in 1983, praising the achievements of the surface fleet, the submarine force and the Royal Marines. But there was not one single mention of the Sea Harrier or its vital success in the air-to-air war. This did not sit well with the Fleet Air Arm. Nor did it reflect Admiral Sir Henry Leach's words, 'Without the Sea Harrier there could have been no Falklands War.'

A final, in-house snub of their aviators by Their Lordships materialised in 1983. It was decided to disband the Naval Staff Directorate of Naval Air Warfare (DNAW) and to give the fixed and rotary wing war-fighters a couple of desks in the Directorate of Naval Warfare. This effectively muzzled the hard-earned expertise of the aviators and of their subsequent influence in the important and often misguided decisions that were to be taken in the future concerning the needs and importance of carrier aviation.

It is a very sad tale of the Naval Staff shooting itself in the foot.

# Chapter 17

# Key UK MoD Decisions Beyond the Falklands

In the 1980s and following the Falklands War, naval aviation was discussed at length within the Naval Staff, and covertly within the Air Staff, but for different reasons.

Having decided misguidedly that the Directorate of Naval Air Warfare (DNAW) should no longer exist, Their Lordships and the Naval Staff appeared to 'tolerate' the fixed-wing Fleet Air Arm but without any full commitment to the importance of the same as a vital part of the Fleet weapon system. (This did not go unnoticed by the Air Staff, who were later to take full advantage of the Naval Staff's grudging support for their fixed-wing aviators.)

The majority of naval captains appointed to command UK's *Invincible*-class carriers were not aviators, and some of these treated their fixed-wing air groups as 'annoyance factors', rather than the vital part of their carrier's air defence. Squadrons had to fit their front-line operational flying training programmes around sporting functions and walkabouts by the ship's company – these were considered to be more important uses of the flight deck.

Nonetheless, and as a direct result of the Falklands air war experience, the Naval Staff did agree to the need to provide the Sea Harrier with a state-of-the-art air intercept weapon system. Thus, the Sea Harrier FA2 was born with its Blue Vixen radar that enabled the full capability of the AMRAAM AIM-120 missile to be realised. This gave the Fleet the long-range, look-down, shoot-down capability that was so sorely missed during Operation Corporate.

Meanwhile, the Air Staff were experiencing serious and embarrassing problems with their Tornado programme. The air defence variant (ADV) had a radar designed by Marconi that just did not work. Air Chief Marshal Sir Keith Williamson had become a director of the company and admitted

Sea Harrier FA2 landing on board.

on television in 1985 that the ADV's radar was no better than concrete ballast. A new radar had to be designed and fitted but proved to be little better. Even with modification after modification to its weapon system, the ADV had no beyond visual range capability until well after the end of the Cold War – and could provide no realistic air defence of the UK homeland. The ground attack and reconnaissance variant also needed serious and expensive modification before it was available to the front line.

Focus on the true availability of the Tornado fleet, or to be precise it's lack of operational capability, was blurred by misleading RAF statements in Parliament that it was fully operational, by the continuing RAF propaganda campaign that it had won the Falklands air war (thereby reducing the military importance of the fixed-wing Fleet Air Arm in the minds of our government ministers), and by anachronistic annual celebrations of the Battle of Britain and the Dambuster raids.

During the 1990s, the Sea Harrier FA2 entered naval service and, with its fully integrated, state-of-the-art, AMRAAM-based air-to-air weapon system, was soon to be applauded by our NATO allies as the most capable fighter interceptor in Europe. Within the UK MoD, no such public approbation or appreciation was forthcoming.

The Strategic Defence Review of 1998 (SDR 98) heralded a possible new way ahead for UK's aircraft carrier operations and air group complements. Thanks to the perseverance and farsightedness of First Sea Lord, Admiral Sir Jock Slater, SDR 98 confirmed the vital need for our carriers to be able to enforce air superiority during expeditionary force operations. This was with the full agreement of the Chief of the Air Staff, Air Chief Marshal Sir Richard Johns. SDR 98 also confirmed the need for our carriers to enjoy a ground attack and reconnaissance capability for the combat air support of ground forces. The Royal Navy VSTOL Sea Harrier FA2 would provide the former and the RAF Harrier GR7/9 would provide the latter. The new fad in the MoD was the expression, 'Jointery' – the conduct of efficient joint operations – and SDR 98 clearly supported this. But unfortunately, these 'best-laid plans' were found to have a one-sided intent.

Following the turn of the century, defence budget constraints led to the debate in the MoD over whether we could afford two different types of VSTOL aircraft on board our carriers. This debate, led by the Air Staff, diverted ministerial attention away from the more pertinent strategic defence issue, 'Why do we have large expensive fleets of non-carrier-capable Tornado and Typhoon aircraft still in service or entering service when their global utility and deployability is virtually zero?'

SDR 98 was ignored, and the VSTOL debate was led by Air Vice Marshal Jock Stirrup RAF and Minister of State in the House of Lords, Lord Bach, who misled the House of Commons Defence Select Committee (in May 2002) – and indirectly the Minister of Defence, James Ingram. They denigrated the now proven capability of the Sea Harrier FA2, claiming that: 'It was not capable of shooting down the main threat to our surface fleet, which are sea-skimming missiles.' De facto, the FA2 weapon system was specifically designed for and capable of dealing with this threat. At the same time, and in total contradiction of SDR 98, they claimed that the embarkation of ground-attack Harriers was more important to task force operations than air defence fighters. The opinion of true naval air warfare experts was not solicited and, shockingly, the Naval Staff under the helmsmanship of First Sea Lord Sir Nigel Essenhigh did not defend the vital need for the Fleet's outer layer of air defence. The decision to withdraw the Sea Harrier FA2 from service was made, based on inaccurate testimony.

Admiral Sir Alan West, later Lord West of Spithead and a strong supporter of fixed-wing naval air power, became First Sea Lord too late to reverse this decision.

For an extended period, the UK Fleet could not now defend itself effectively against long-range air and missile attack throughout the global commons. Against all military logic, a vital element of the Fleet's strategic capability had been removed. But even now in 2024, the legacy of the Sea Harrier lives on, as shown by Matthew Houghton on the 'We like Harriers' page on Facebook. He presented the following tribute:

> The Sea Harrier.
> In the vast blue sky, where dreams take flight,
> A legend was born, with wings of might,
> The Sea Harrier, a marvel in the air,
> A symbol of triumph, beyond compare.
>
> Oh, Sea Harrier, conqueror of the sky,
> With grace and power, you soar up high,
> Through battles fought, you never tire,
> A hero of the sea, your success inspires.
>
> From the decks of carriers, you took your stand,
> A warrior of the sea, protecting our land,
> With vertical take-off, you'd rise with pride,
> Ready to defend, with courage by your side.
>
> Through stormy weather, you'd navigate,
> With radar precision, you'd never hesitate,
> Your missiles locked on, enemies beware,
> The Sea Harrier strikes, with deadly flair.
>
> From the Falklands to the Persian Gulf,
> You proved your worth, with every pulse,
> Engaging the enemy, with skill and might,
> The Sea Harrier shining, in the darkest night.

Oh, Sea Harrier, conqueror of the sky,
With grace and power, you soar up high,
Through battles fought, you never tire,
A hero of the sea, your success inspires.

Now retired, but your legacy remains,
A symbol of triumph, forever ingrained,
The Sea Harrier, a legend we'll recall,
For its triumphs and victories, standing tall.

Oh, Sea Harrier, your story we'll tell,
Of bravery and triumph, where legends dwell,
Forever remembered, in the annals of time,
The Sea Harrier's success, an eternal rhyme.

In times of high tension or combat, the Royal Navy had to shelter under the air defence umbrella provided by United States Navy strike carriers. Not long after the withdrawal of the Sea Harrier from service, and during the Afghanistan crisis, the RAF and naval strike wing Harriers were withdrawn summarily from service and without notice at the behest of Air Chief Marshal Sir Jock Stirrup, who persuaded Prime Minister David Cameron, to ignore National Security Council recommendations that the obsolete Tornado GR4 should be withdrawn instead. (The cost of retaining the GR4 in service was estimated to be £7 billion, whereas retaining the relatively new and capable Harrier was estimated to be just £1 billion. There was no financial or operational justification for ignoring the Security Council's guidance.) Stirrup and the RAF had achieved their aim – taking away all the fighter and fighter bomber aircraft of the fixed-wing Fleet Air Arm.

However, during this persistent emasculation of Royal Navy aircraft carrier power and after much infighting within the MoD, the build of two new strike carriers for the Royal Navy was approved. But their final configuration was to prevent them enjoying full strike carrier capability.

This was a direct result of the much-acclaimed policy of 'Jointery'. Senior officers from the RAF in particular (but including at least one army brigadier), without any appropriate naval air warfare experience or expertise, were given control of the 'Carrier Strike Project'. This led

Sun sets on the forsaken legend – Sea Harrier.

to our two new capital warships, the *Queen Elizabeth* class of aircraft carrier, being fitted with a ski jump on the flight deck instead of an angled deck with catapults and arrester gear. This severely limited the choice of fixed-wing aircraft that could be embarked in the new warships. The only aircraft capable of operations from the flight deck was the short take-off and vertical landing (STOVL) F-35B. Defence suppression aircraft (EA-18G Growler), Hawkeye airborne early warning (AEW) aircraft and air-to-air refuelling (AAR) aircraft could not operate from the deck.

The F-35B Lightning II stealth fighter is fully capable of providing the outer ring of air defence for the Fleet with its state-of-the-art hard-kill and soft-kill weapon system, but this capability can only be realised if enough F-35B aircraft are procured to enable a 24/7 capability – which is vital for modern naval air warfare combat operations. To date, adequate procurement numbers have not been ordered.

A viable and logical solution to the limitations facing our capital warships would be to reconfigure them with an angled deck, catapults and arrester gear, thereby allowing the embarkation of a fully capable Strike carrier air group.

The ongoing 2023 House of Commons Defence Select Committee Inquiry into armed forces readiness will hopefully address these issues. The published witness statement to that inquiry by 'Sharkey' Ward remains under consideration and reads as follows.

Two F-35B Lightning fighters over HMS *Queen Elizabeth.*

**'Written evidence submitted by Commander N.D. MacCartan-Ward DSC AFC**

i. This inquiry provides the much-needed opportunity to address the *raison d'être*, constitution and mobility of our individual armed services within the context of being able to protect the UK homeland base and to project military power rapidly and effectively in response to emergent crises in the wider region.

ii. This submission addresses these factors and, in doing so, suggests that:
   a. The 'hollowing out' of the Army with its heavy armour weapons systems is justified due to their lack of strategic mobility on the global stage.
   b. The 'hollowing out' of the Royal Navy over recent decades has been unjustified in the context of declared strategic maritime policy and needs to be reversed urgently if we are to be considered a tier One fighting force that is globally capable of providing effective support to our allies.
   c. There is justification for the 'hollowing out' of the UK's large combat fleet of land-based tactical fighter aircraft due to their inability to protect the UK homeland base from long-range air attack and their lack of strategic mobility in strength. The Future Combat Air System (FCAS), the Tempest Project, is the latest 'Cuckoo's egg to be laid in the UK's defence procurement nest'. The Committee may wish to demand a full review of the formal Ministry of Defence Requirement document underpinning the Tempest project. The specification there-in should explicitly require the Tempest aircraft to be fully capable of operating from our carriers – thereby giving it the necessary strategic utility. If that is not the case, this collaborative project must be suspended until a valid formal requirement is forthcoming.

**Introduction**

1. Military logic and thinking-outside-the-box of 'traditional wisdom' must be applied when addressing the areas of interest iterated by the Committee. The strategic *raison d'être*, mobility and utility of each

armed force must be scrutinised. We can then assess its ability or otherwise to respond to crises effectively at a moment's notice. This capability, or lack of it, may then be balanced against the principal perceived threats to our global commercial, trade and energy supplies upon which our prosperity and that of our allies depends.

2. The UK in conjunction with our NATO and other allies must be able to project sufficient flexible military maritime power throughout the global commons to deter and, if necessary, respond to those that would harm our offshore national interests, threaten the security of our homeland base and/or threaten the safety of our overseas territories. The ability or otherwise to project such power should be the fundamental guideline to the constitution of and investment in weapons platforms and systems for each armed force.

**Assumptions**

3. In any direct confrontation with Russia, the principal conventional threat to the UK would be long-range, air-to-surface missiles launched primarily by waves of Russian strategic bombers rounding the North Cape.

4. Investing in the capability to project robust military maritime power throughout the global commons serves two purposes:
   - It strengthens the UK's ability to provide realistic protection of the UK homeland base from air and sea attack.
   - It demonstrates a significant UK contribution to the global defence of the offshore interests of fellow NATO partners and our allies.

**Are the armed forces sufficiently capable, resourced and ready to protect the UK and our allies? What are the main gaps in capability and/or readiness, and what will it take to fill these gaps?**

**Protecting the UK Homeland Base and our Allies: The Royal Navy**

5. Protecting the UK homeland base and our allies from direct attack can best be assured by adherence to measured investment in weapons

platforms and systems that directly support declared strategic maritime policy. This is part of the *raison d'être* and mission of our Navy, which includes working with our allies utilising mobility, flexibility and visible deployed power to ensure the safe passage of our trade and energy supplies throughout the global commons.

6. A well-armed, well-resourced Navy would have the ability to deploy a Carrier Battle Group and its air defence weapons systems up threat from the UK and provide a 24/7 defensive barrier against, for example, Russian missile-armed strategic bombers that would otherwise be able to approach our shores with impunity to within the delivery range of their long-range missiles (1,200 nautical miles). It would also add weight to the deterrence and combat power of the joint Allied initiative in the Indo-Pacific region.

7. Unfortunately, over the past few decades our Navy has suffered a Draconian reduction in size – a 'hollowing out' that does not reflect the UK's declared strategic maritime policy or the perceived global threats that we now face.

8. Our global commitments and alliances represent an urgent need for a larger Fleet with well-armed weapons systems beneath the sea surface, on the surface and above the surface. The introduction to service of our two new carriers is a step in the right direction. But failing to provide them with their own robust complement of F-35B STOVL fighter aircraft (their principal offensive and defensive weapon system) denies them the full capability for rapid and flexible response to emergent crises whether in the Atlantic, the Middle East or the Indo-Pacific region.

9. This shortfall merits immediate correction.

**Protecting the UK Homeland Base and our Allies: The Army**

10. Currently, our standing Army does not appear to have a *raison d'être* that is consistent with:

a. a robust capability to defend the UK homeland base, or with
b. declared national strategic policy.

11. For example, it is not configured with defensive surface-to-air missile systems such as Iron Dome and Patriot that could intercept Russian long-range air-to-surface missiles targeting key installations such as our military airfields, naval bases, ports, power stations, etcetera. The initial lack of air defence missile systems in Ukraine and the carnage it led to is worthy of note.

12. Our island nation infrastructure is particularly susceptible to such attack.

13. It is suggested therefore that investment in proven surface-to-air defensive weapon systems should now take the place of the current major investment in heavy tanks, artillery and armoured personnel carriers that have:
a. Zero capability against this Russian threat, and
b. Little or no strategic mobility.

14. This suggestion does not imply any failure to support the NATO alliance. On the contrary, it recognises the need for the UK, individual NATO members and other allies to address their unique geographical, territorial and sovereign status and to plan their military investment and resources accordingly. A large standing army of 'boots on the ground' and a well-funded tactical air force would appear to be a national security priority for those states enjoying a territorial border close to or with Russia. That is not the case with the UK.

15. The UK, as an island nation and far from the eastern front, should not commit significant amounts of our defence budget on bolstering the capability of these landlocked NATO allies through the maintenance of heavily armoured units and a large standing army that cannot be easily and rapidly deployed to the eastern front or beyond in times of crisis. The Ukraine war has demonstrated how

difficult and time-consuming is such deployment (even from within the European mainland).

16. It follows on from paragraphs 10 to 15, above, that we in this island nation, geographically distant from the Russian front, would be less than wise to invest our limited financial resources in substantial land-based weapons platforms and systems that are not strategically mobile enough to respond rapidly, if at all, to crises away from our homeland base.

17. The UK's strategically mobile 'boots on the ground' capability is vested almost entirely in the Royal Marines, and they should therefore receive appropriate investment.

**Protecting the UK Homeland Base and our Allies: The Royal Air Force**

18. Some of the land-based strategic bombers of the US Air Force are designated for rapid deployment to the UK in times of tension in order to provide a capability that is beyond the financial reach of the UK's defence budget. They supplant any associated strategic *raison d'être* that might be claimed by the RAF, who can only aspire to being a tactical as opposed to a strategic air force – without the latter's reach and deployability in strength during global crises.

19. The UK's only strategic (as opposed to tactical) fighter air defence and offence capability is vested in the F-35B air groups planned for embarkation in our two aircraft carriers. This strategic capability is currently and seriously under-funded and under-resourced.

20. The Committee might wish to note a lack of investment balance within the NATO alliance regarding the procurement of short-range tactical fighter aircraft. Why did Britain's RAF initially order 232 Typhoon tactical fighter jets (although eventually only acquiring 160) while the more economically powerful German nation bordering on Russia at first ordered fewer at 180?

21. The lack of strategic utility and the hundreds of billions of pounds invested in the Typhoon and its predecessor Tornado (more than 300 procured) represent a veritable 'cuckoo in the nest of other properly justified military procurement programmes'. Today, the multi-billion-pound Future Combat Air System (FCAS), the Tempest project, is the latest cuckoo's egg to be laid in that nest. This collaborative project has been conceived without any transparent military justification – and the secrecy surrounding its design specifications indicates that it will have no more strategic capability and utility than its non-carrier-capable Tornado and Typhoon predecessors: a serious waste of defence budget funds.

22. In the light of the significance of para 21, above, the Committee may wish to demand a full in-house review of the formal Ministry of Defence Requirement document underpinning the Tempest project, if there is one. In all logic, this should be a 'Naval and Air Staff Requirement (NASR)' and the specification there-in should explicitly require the Tempest aircraft to be fully capable of operating from our carriers – thereby giving it the necessary strategic capability.

23. As intimated at paragraphs 5 to 8, above, the UK's land-based tactical fighter aircraft do not have the range and endurance on task to provide an effective 24/7 air defence barrier that would deter or prevent Russian strategic bombers from launching their air-to-surface missiles at UK targets (the principal *raison d'être* of the RAF*)*. Those who claim that its fighter aircraft radius of action can be readily increased using air-to-air refuelling (AAR) are deliberately misleading the Government.

24. The UK's now limited AAR capability was seriously downgraded with the introduction of the £13 billion Voyager project – a small number (12) of ultra-expensive aircraft, of which less than half a dozen would be available for global and local AAR military tasks.

25. To provide an effective 24/7 barrier against the Russian air threat with its long-range (1,200 nautical miles) missiles in times of high

tension, land-based tactical fighter aircraft would need to maintain a robust combat air patrol presence (two pairs of Typhoons) at least 1,000 nautical miles from the UK base. Each Typhoon would have a transit time to and from its CAP station of two hours each way and a time on task of two hours – a six-hour flight time for each aircraft at a cost of £75,000 per hour, equalling £450,000 per aircraft sortie, and £1.8 million per CAP cycle (£75,000 ×6 ×2 ×2). Twelve rotations of the CAP each 24 hours would cost £21.6 million per day. Every 14-day period of high tension would therefore cost £302.4 million – and for eight weeks of high tension, approximately £1.2 billion. Such costs are prohibitive and do not include the vital AAR support for making a 24/7 barrier possible. Needless to say, UK does not have nearly enough AAR capability or enough Typhoon pilots to generate and sustain an effective barrier.

26. Carrier battle groups, with F-35B fighter aircraft embarked, provide the cost-effective answer to this long-range air defence need, as realised and provided successfully throughout the Cold War with the Soviets.

27. In the light of all the above, it is evident that the main *raison d'être* of the RAF, i.e. the robust air defence of the UK homeland base, cannot be fulfilled with its collection of relatively short-range, land-based tactical fighter aircraft. Hence the need to divert funding to the balanced implementation of the UK's Strategic Maritime policy.

**Are the UK armed forces a 'tier one fighting force'? Do they need to be?**

28. The quality of our military personnel, soldiers, sailors and airmen remains equivalent to that expected of a tier one fighting force. But they are not being provided with adequate weapons platforms and systems that ensure the mobility and flexibility needed for the UK to respond effectively and urgently to emergent crises, whether in conjunction with or independently from our trusted allies.

29. To date, our Whitehall masters have:
    a. Paid lip service to our declared strategic maritime policy, centred as it is on our two new aircraft carriers and their associated battle groups (including Royal Fleet Auxiliary logistics support). Instead, they have overseen the 'hollowing out' of our Royal Navy Fleet upon which that policy depends.
    b. Perpetuated the myth that UK needs a large standing army supported by heavy armour and have invested in associated weapons systems that have zero strategic mobility and, therefore, a much-diminished *raison d'être* and mission (resulting in a justifiable 'hollowing out' procedure that is not yet complete).
    c. Invested misguidedly (and continue to do so at major cost – including the militarily unjustifiable Tempest project) in large fleets of relatively short range, land-based tactical fighter aircraft that have:
        i. Little or no global capability to respond rapidly and in strength to emergent crises outside the NATO area, and
        ii. Too limited radii of action to be able to provide a robust 24/7 air defence of the UK homeland base.

30. With respect to the independent/unilateral global projection of military power, it is reasonable to accept that the UK is not equipped to be a tier one fighting force. However, if properly equipped and working in close conjunction with our allies (particularly the US Navy), the Royal Navy can and would provide a tier one contribution to the policing and stability of the global commons.

**What are the consequences of the army having been 'hollowed out and underfunded'?**

31. The 'hollowing out and underfunding' of the Army is testament to the fact that maintenance of a large boots on the ground standing army supported by expensive heavy armour weapons systems, neither of which can be deployed rapidly offshore in times of crisis, and which therefore have little if any strategic relevance, has become an anachronistic burden to our national security initiative.

When complete, it should have released important defence budget funding for the implementation of the UK's declared strategic maritime policy and the measured expansion of our Fleet weapons system capability. As such, this 'hollowing out' would be of direct benefit to our national security.

32. The Committee will wish to recognise that the consequence of the recent 'hollowing out and underfunding' of the Royal Navy, on which the realisation of our declared policy depends, has diminished the UK's ability to protect the UK and our allies and to deter those that might otherwise harm our joint global interests. It is this that has brought into question whether the UK continues to be a tier one military force.

**Which of these consequences needs to be addressed most urgently?**

33. The 'hollowing out and underfunding' of the Royal Navy needs to be addressed and reversed as a priority in support of and in keeping with the UK's declared strategic maritime policy and the global threats that now face us.

34. The 'hollowing out' of the UK's standing Army needs to continue, and its modern *raison d'être* established. Its anachronistic investment in heavy armour weapons systems is not representative of any utility or capability to respond urgently to emergent crises offshore. Nor do these weapons systems have any relevance to the desired robust protection of the UK homeland base.

35. The *raison d'être* of and the major investment in the RAF's tactical fighter force needs to come under detailed scrutiny. It must be 'hollowed out' to reflect:

    a. Its limited capability to protect the UK base from sophisticated long-range air attack.
    b. Its inability to be deployed rapidly in strength when emergent global crises so demand.

**Are the Government's plans sufficient to address any shortfalls?**

36. Current Government procurement plans:
    a. Do not support the declared strategic maritime policy adequately.
    b. Do not recognise the inability of the Army and the Royal Air Force to respond rapidly and with strength to emergent crises throughout the global commons (and to some degree within Europe). Their ability to contribute to the effective and timely projection of British political and military power and influence is limited.
    c. Appear to sacrifice urgent military need in favour of providing lucrative weapons systems contracts to the UK's defence contractors.

37. The lack of urgency and commitment to build a stronger Navy is misguided in that it reduces the UK's ability:
    a. To provide robust protection of the UK homeland base, and
    b. To project military power when reacting to instability in the wider region.

13 May 2023

'Lest we forget our maritime history'. Clean-shaven Sharkey in Grenada does remember.

# Part III

# Beyond the Falklands War

# Chapter 18

# The Continuity, Impact and Significance of the USAF Presence in the UK

There is a poignant scene at the beginning of the 1949 movie *Twelve O'Clock High* when the Adjutant of the fictitious 918th Bomb Group returns from the United States to his former UK airfield. Played by Dean Jagger, the movie character Major Stovall (promoted in the movie to lieutenant colonel, US Army Air Force) goes back and stands at the end of the runway and thinks back on all that happened in the Second World War to his bomb group. It is a powerful movie. Jagger won an Oscar for best supporting role.

The movie exemplifies the leadership and sacrifices made by the Eighth Air Force and one bomb group. The movie could well have been about the 100th Bomb Group.

After the Second World War, the 100th was not forgotten and a group of people organised and raised the necessary funds as charitable donations to create an extraordinary memorial by building a museum at the former air base of Thorpe Abbots in Diss, Norfolk, that contains the history and many of the combat records of the group's outstanding achievements in helping win the war.

The recognition by one community in England of one US wartime bomb group also has wider significance. The continuity of a US Air Force presence in the UK persists to this day and has equal tactical and strategic significance as in the Second World War. The USAF's presence in the 2020s opens the door for the UK to re-evaluate its plans, policies, and operations at the highest level.

Let us look at where the USAF is today in the UK and where it is headed for the foreseeable future with its closest ally.

Before leaving the movie, it is important to note that in 1998 *Twelve O'Clock High* was selected for preservation in the United States National Film Registry by the Library of Congress as being, 'culturally, historically, and aesthetically significant'. The main characters in the movie were modelled on real Eighth Air Force personnel. For example, Millard Mitchell plays Lieutenant General Pritchard, modelled on the Eighth Air Force's first Commander, Major General Ira C. Eaker.

The title of the movie reflects the Eighth Air Force's calling out the positions of attacking enemy aircraft with reference to an imaginary clock face, with the B-17 at the centre. So 'twelve o'clock high' meant the enemy was directly ahead and above.

Since the Second World War the United States Air Force (created in 1947 at the same time that the Central Intelligence Agency was founded alongside other major legislative changes that saw the founding of the US Department of Defense and Chiefs of Staff organisation) has had a major presence in the United Kingdom. This was referenced earlier with a description of the various bases and facilities, covering the complete spectrum of air and space warfare, supported by considerable fully networked intelligence and information warfare capabilities and facilities.

Within both special highly classified US-UK links and agreements, and the wider Five Eyes Intelligence associations with Canada, Australia, and New Zealand, the US Air Force has had and still maintains an extraordinary continuity of commitment to a global strategy that requires today and for the foreseeable future the continuing presence of highly significant US Air Force capabilities in the UK.

This USAF presence and capabilities must be reviewed alongside UK current and future UK plans, policies, and operations, and in particular the UK's acquisition strategy in a country of a vital US ally that does not have the resource base of the United States. The UK must choose wisely in a current and projected threat environment in which procurement decisions should not be taken without examining the full panoply of USAF resources in the UK.

A review of any of the reputable and reliable open sources of USAF current capabilities, such as *Janes All the World's Aircraft* 2022/2023, and the various US Air Force Association publications, together with the public US Congressional records on annual appropriations by the House of Representatives and Senate, reveals the huge disparity between the Order of Battle of the US Air Force and the Royal Air Force. The United States has the resources to invest billions of dollars in new systems and technologies that the UK cannot contemplate. The corollary to this is that given the incredible investment made by the US in 'special programmes' that in due course become public knowledge and are revealed as new inventory for the USAF, the UK cannot match in any shape or form the air systems that are required to ensure air and space superiority in the current and emerging threat environment. The costs are beyond the budget limits of the UK.

There are several key ways and means by which the UK can continue to not only contribute in the scientific and technological domains to the significant benefit of the UK industrial base and at the same time enable the UK to be fully integrated in an alliance-based deterrent strategy that is both affordable and achievable by the UK.

The context for the above should first be stated because it is the overarching strategic concepts and plans that drive the acquisition and investment policies and plans of both the US and the UK.

What are these and how will they shape the future?

World events in 2022 have set the stage. In many places there is not just discord and disharmony but also bloodshed on a large scale. The war in Ukraine has focused attention daily on a people suffering from the effects of Russian tyranny the likes of which we have not seen since the evil days of the Third Reich. Putin is not just a despotic demagogue. He is also a master of deception and lies to a people whose communications and freedom to think and express their opposition are totally controlled.

The common bonds and processes that unite the US and the UK and other key allies in facing the current threats are critical in creating a sound strategy. One key word encapsulates such a strategy: 'deterrence'. From the beginning of the Cold War and the formation of the NATO alliance, the US and the UK and their allies succeeded in keeping the world safe from Armageddon, including staving off the Soviet Union's nuclear militarisation of Cuba in 1962. President Kennedy and his highly capable body of advisors used naval deterrence to make it patently clear to Nikita Khrushchev's Soviet Union that deploying nuclear armed missiles in Cuba was a non-starter. The disposition and power of the United States Navy came through. The President's team of Robert McNamara, General Maxwell Taylor, General Curtis LeMay, Admiral George W. Anderson, and the Attorney General Robert F. Kennedy stood firm and refused to be intimidated and bullied by their opposite numbers in Moscow, particularly Khrushchev's closest cohorts, Anastas Mikoyan, Rodion Malinovsky, and Matvei Zakharov. Admiral Anderson, the Chief of Naval Operations, was in charge of the naval blockade of Cuba.

Sadly, Anderson never became Chairman of the Joint Chiefs of Staff, having differences with the difficult and controversial Secretary of Defense Robert McNamara. President Kennedy subsequently made Anderson Ambassador to Portugal, and later he became President of the Foreign Intelligence Advisory Board. *Time* magazine featured Admiral Anderson on its front cover and described him as 'An aggressive blue-water sailor of unfaltering competence and uncommon flare.'

The point is quintessential from the above 1962 scenario: 'deterrence works'.

Perhaps sadly, this tradition and *modus operandi* has not been continued in the modern era. Let us look at two critical events, one involving China, and the other involving Russia. In the case of the former, the US allowed the Chinese to unilaterally take over and seize key islands in the South China Sea in total contravention of international law, and the rulings of the International Court of Arbitration. The rights of several nations were violated in the process, including the Philippines, Taiwan, Malaysia, Vietnam, and Brunei. The United Nations Convention on the Law of the Sea (UNCLOS) was swept aside by the Chinese in blatant naval actions against the interests of these nations and the overall security posture in East Asia. In addition to the militarisation of these islands, the Chinese were clearly and demonstrably looking to secure access to the undersea oil and mineral deposits off these islands. UNCLOS provides a 200-mile economic boundary around the coastline of signatory nations. This is a non-trivial area of ocean.

The United States, with a powerful and highly capable Seventh Fleet, within the US Pacific Fleet, stood by and let the Chinese get away with all this. If this had been the Cold War, there is no way that any American President and its allies would have allowed this kind of unilateral seizure of others' sovereign territory that had, in addition, serious strategic implications for the future of the United States' and Allied interests. The Chinese could have been signalled well in advance that such moves would not be acceptable, and the US could have implemented a major blockade with a powerful US Navy and Marine Corps force. This is not confrontational. This is the wise and accepted, in historical terms, deterrent value of maritime forces signalling that outright aggression will not work. Are there risks? Yes. However, the alternatives, if ignored, are even more risky. Deterrence works if executed in a timely manner with full force that is credible and clearly capable of protecting, in a military sense, the vital national interests of the United States and its allies. What we did was to embolden the Chinese, to give them encouragement to be more aggressive.

The year 2022 was the 100th anniversary of the death of the British naval strategist Sir Julian Corbett. In comparative terms Corbett is the UK's equivalent to the United States' great nineteenth-century strategist Alfred Thayer Mahan (1840–1914, aged 74). The work of both these fine men, with strong historical references in their works, exemplifies in part the need to reassess, not necessarily in complete Mahan-Corbett terms, the roles of deterrence in modern strategy.

The relevance of the past for the present and the foreseeable future, in the context of a 'deterrence strategy' may be stated as the use of power in all its forms in the pursuit of peace and the vital national interests of the United States, the United Kingdom, and their allies, and this entails the wise and unequivocally clear use of military power, coupled where appropriate with soft power and sanctions. The latter have to be measured and well thought through in terms of overall consequences. Diplomacy must always go hand in glove with all these measures though not in ways that provide belligerents with time to stall and prepare for outright aggression. In the current and future global security environment the United States is pursuing a global maritime strategy closely interwoven with air and land power, while maintaining traditional and very necessary land-based deterrent strategies that require extensive and highly capable air support capabilities and operations. The European, Baltic, eastern Mediterranean, and Middle East theatres are critical parts of the strategic posture of the US Air Force. The deterrent posture of US and Allied resources in the Indo-Pacific Command of the United

States illustrates how well organised and regularly exercised combined forces can demonstrate resolve and resilience in the face of threats and provocation.

If resolve and resilience underwritten by effective capabilities are not demonstrated the worst case can quickly occur. In 2014 Vladimir Putin unilaterally took over the Crimea, in total violation of international law, and also shamefully ignoring the 1994 Budapest Memorandum that the United States, Russia, and Britain agreed to and 'committed to respect the independence and sovereignty and the existing borders of Ukraine', and (most important) 'to refrain from the threat or use of force' against Ukraine. The situation faced in 2022 resulted in part from a failure to 'deter'. Proactive deterrent measures under the Budapest Memorandum could have been taken to checkmate Putin.

Prior illustrious US and UK leaders such as Kennedy, Roosevelt, Churchill and Thatcher understood the application of strength, courage, resilience, and clearness of purpose in facing formidable enemies. In the current and projected international security environment the leadership of the US and UK and key allies requires clearness of purpose to harness an effective 'deterrent strategy'. The latter requires a wise application of the UK's more limited defence resources, considerable nonetheless, though not comparable with the US. The balance of resources and application of UK power in the pursuit of peace alongside the US now requires an urgent look at how the USAF presence and capabilities in the UK may enable the UK to concentrate its resources in optimal ways.

Let us look at the current and future significance of the USAF presence in the UK and how this may enable the UK to both plan and make critical defence investments in the context of an allied strategy that melds with the combined vital national security interests of both nations.

The continuity, significance, and impact of the US Air Force's extensive presence in the UK now require careful review. The relevance for current and future British defence policy across the whole spectrum of activities, driven by a top-down break-down based on the strategic deterrence case made earlier for alliance-based cooperation, will be analysed.

One key initial factor is the relative annual defence investment made by the US and the UK. There is a huge disparity, even though the UK more than carries its fair share of the NATO defence burden financially and at the operational level. The US defence budget request in fiscal year 2022 rose to $722 billion, an increase of $17 billion from 2021. The Biden Administration sent a budget request of $737 billion for the Department of Defense for fiscal year 2023. The 2022 fiscal year was broken down as: Army $173 billion; Navy $212 billion; Air Force $213 billion. The Department of Defense planned to spend $72.7 billion on aircraft alone; ships and submarines $39.3 billion. By contrast the UK's defence budget was $59.2 billion for 2022. Germany's was $52.8 billion, and France's $52.7 billion. Japan spent $49.1 billion, South Korea $45.7 billion, and Russia spent $61.7 billion.

In Fiscal year 2023 the US Air Force had a requested budget of about $234.12 billion. This included the Space Force. The enacted budget for the US Air Force in fiscal year 2022 was $222.28 billion. These two numbers should be reviewed apropos of the total UK defence budget of $59.2 billion in fiscal year 2022.

Given the UK's ongoing financial difficulties it is reasonable to reassess what may or may not be possible for Royal Air Force costs, investments and key acquisitions. The RAF equipment plan is ambitious over the next ten years plus. The current financial plan may say one thing; the downstream reality may be different in a fiscally challenged country. The $49 billion earmarked for investment in the acquisition of new equipment may or may not materialise. The motivation is highly laudable with key statements about making the RAF technologically innovative, productive, and a lethal air force. Modernisation of the RAF's Typhoon force and acquisition of a Future Combat Air System, the Tempest, are top of the RAF's list. No UK operational justification has yet been revealed for each of these two extremely expensive tactical aircraft programmes that devour an unwarrantedly large percentage of defence funding that could be better spent on other more strategically capable military programmes. However, the investment numbers for these and unmanned programmes are simply tiny in comparison with current USAF investments, programmes, let alone the highly classified 'special programmes' budget that does not see the light of day and is not in the public arena. The latest public revelation of a key new system emerging from USAF classified programmes will be discussed later.

What the above reveals leads to the asking of fundamental questions about UK defence investment plans in light of the overarching deterrence strategy discussed earlier and the relatively huge differences between USA and UK defence budgets.

The continuity of the USAF's presence in the UK is a given. The significance is huge and has to be recognised by the British government and the Ministry of Defence.

The US is moving ahead in air warfare superiority domains where the UK simply does not have the financial ability to contribute let alone run parallel programmes. One domain illustrates this point. Long-range airborne refuelling using unmanned vehicles has already been created and proven in the United States. For example, the MQ-25 Stingray provides on demand air-to-air refuelling permitting US carrier-based aircraft to extend their combat ranges beyond existing maximum ranges, thereby permitting 'deep strike' capability. The Stingray has a 500 nautical mile range, at least 15,000lb of fuel for air-to-air refuelling and low-observable features. This type of system helps the USAF with its tanker shortage in its fleet of KC-135 Stratotankers and KC-10 Extenders. The new KC-46, intended to bridge the USAF tanker gap, can be augmented by unmanned refuelling vehicles. In true American defence competitive vein, the Lockheed-Martin Company has produced an alternate based on the Airbus A330, a direct competitor to the above Boeing KC-46, based on Boeing's 767 airframe.

In this environment the Royal Air Force can contribute to USAF UK-based aircraft by a plan that more extensively integrates RAF refuelling capabilities with the operational needs of UK-based USAF fighter and bomber aircraft. This role is hugely significant in operational, tactical, and strategic terms.

From a naval perspective the Stingray will enable Royal Navy and Royal Air Force carrier-based aircraft to enter contested airspaces at very long ranges from the UK's two aircraft carriers, HMS *Queen Elizabeth* and HMS *Prince of Wales*.

The RAF can complement the USAF's presence in the UK in other valuable domains at a junction in air warfare history where the USAF simply has dominance in terms of both sheer numbers and current and future capabilities. Cyber operations, both defence and offence, are a very good example.

The impact of cyber-attacks can be massive if not anticipated and countered before taking effect. In the civilian commercial domain, the 2017 'NOTPETYA' cyber-attack on the Maersk shipping line saw a highly organised and well-led company lose $500 million in two minutes when its terminals were attacked, losing more money than the total effects of Somali pirate attacks from 2005 to 2015; two minutes of cyber-attack versus ten years of piracy.

The RAF can support USAF cyber operations in conjunction with other sensitive US-UK intelligence operations based in the UK. Information warfare in support of the USAF can become a dominant role for a future Royal Air Force. The latter can combine its resources and capabilities with those of the British GCHQ and the wider Five Eyes Intelligence community.

At the air warfare operational level, the USAF celebrated its seventy-fifth birthday on 18 September 2022 (founded by Act of Congress in 1947 as a successor to the US Army Air Force) by many demonstrations of its ability to conduct major operations from the UK. The USAF also demonstrated how quickly and effectively it can augment UK-based capabilities. In August 2022, B-52 strategic bombers landed at RAF Fairford. They flew non-stop from Minot Air Force Base in North Dakota. Minot is the home of the 5th Bomb Wing and 91st Missile Wing, both part of the USAF's Global Strike Command. The B-52H Stratofortress is the key mainstay of the 5th Bomb Wing. The legacy with the Eighth Air Force and the 100th Bomb Group in the UK in the Second World War becomes very clear. The USAF can augment its capabilities and operations in the UK from the continental United States very rapidly indeed.

The continuity and significance of the role of US new and innovative airborne and space systems is very well illustrated by the recent revelation by the USAF of its latest platform to emerge from the classified world. This platform will undoubtedly be deployed to the UK in due course. There is no way that the UK and the Royal Air Force could ever create such a capability. The costs are simply too exorbitant.

The new USAF B-21 bomber was officially launched on 2 December 2022, at Northrop Grumman's Palmdale, California, plant. There was, naturally, very tightly controlled viewing.

At the strategic level, and specifically in terms of US-UK relations, the B-21 is the lineal successor to the B-17s that were described earlier regarding wartime Eighth Air Force operations. Nothing has changed. Time and technology have marched on. However, relations have not changed, and the strength of US-UK relations are as enduring today as they were in the darkest days of the Second World War. The key difference is that the UK must decide on how it fits its more limited resource base into the US-UK and allied strategic deterrence tapestry.

To continue a regular Air Force that is increasingly out of date with limited, if any, global strategic applications, is now a major question for review. The new USAF B-21 and a multitude of other capabilities requires the UK to re-examine its priorities in ways that fit the new global strategic threat environment.

Hypersonic weapons are increasingly becoming game changers. US space and USAF assets, together with US Navy and US Marine Corps capabilities, have to be reviewed by the UK in terms of what it can contribute sensibly in both a technical and financial sense. What the UK contributes must be relevant, meaningful in terms of alliance relations, and achievable technically and financially. If the UK reviews USAF operations from the UK over the past several decades, it becomes very clear what a way ahead should entail. To mention just a few achievements, from the Berlin Airlift to Operation Desert Storm in 1991, to Operation Allied Force to stop the Serbian slaughter of ethnic Albanians in Kosovo in 1999, the USAF demonstrated how UK-based assets could challenge, confound, and deter threats. Today the Indo-Pacific theatre of operations offers challenges. US General Kenneth Wilsbach advocates allied togetherness and military cooperation across the board. This theatre is one in which the Royal Navy and the Royal Marines play an increasingly significant role. This is a natural lead in to where the UK should critically examine its current and future force structure, investment, and deployments in pursuit of an alliance policy of a deterrent posture based on power in the pursuit of peace.

Before moving on to the above issue there is one key domain where the UK must re-examine its role in relation to the United States and specifically the roles and missions of the USAF in the UK. This is the future of the UK industrial base.

The UK has very well-established defence acquisition programmes with the US. In electronic warfare alone British Aerospace Systems (BAeS) has massive investment and business in both the US and the UK. The BAe EC-130H Compass Call programme is just one example. The EC-37B future system will be a ground-breaking capability. BAeS contributes significantly to the F-35, F-35B, F-15, and F-22 programmes. RAF refuelling capabilities were mentioned earlier. Rolls-Royce supports the USAF with multiple innovations and has a significant business base. The USAF C-130J Super Hercules airlifters are powered by Rolls-Royce AE-2100D3 turboprop engines. Most significant of all is that the whole of the USAF B-52 fleet will be re-engined with Rolls-Royce F130 engines, with much increased fuel efficiency and better and easier maintenance for USAF technicians.

The above illustrates that the UK industrial base will not be challenged negatively in a restructuring of its defence posture.

The continuity, impact, and significance of the USAF's presence now and in the future in the UK has been made.

The major question now is, simply, what does the UK do in terms of reorganising its overall defence posture, current and future programmes in light of all the above?

# Chapter 19

# The Strategic Drivers that Should Effect Change

Before analysing and then concluding what is the best way ahead for the United Kingdom's defence posture across the complete panoply of relevant issues, it is critical to be sure of one key issue. What are the strategic drivers that currently and into the foreseeable future drive change in the UK?

The current publicly available threat assessments that are available in the media speak for themselves. Open-source testimony in the United States House of Representatives and Senate Select Committees on Defense and Intelligence, together with the appropriate Appropriations Committees, provide highly reliable information. The latter are equally important because their output reflects what the United States will invest in response to the various global threats.

The various key global threats are well known and need no detailed elaboration.

China presents the greatest threat at multiple levels, not simply the Taiwan invasion issue. Leaving aside China's expansionist military activities, China's control and distribution of critical precious metals for the software and defence industries makes it necessary for the US and the UK and their allies to have alternative sources. Taiwan's information and telecommunications industries are prodigious. The latter may be categorised as an understatement. US and UK dependence on key Taiwanese robust advanced semiconductor manufacturing capacity is at the heart of Taiwan's independence. Taiwan's Ministry of Digital Affairs (MODA) says it all in its various publications. It would be a disaster from purely a digital supply point of view for Taiwan to be invaded and China gain control of Taiwan's critical global software, computer services, and critical next-generation telecommunications products.

Leaving China, next is Putin's Russia and the war in Ukraine says it all. What Putin and his coterie of inner group supporters will do next remains to be seen. What we do know is that Putin only understands one thing: countervailing deterrence that is underwritten by a clear and unequivocal strong alliance of nations who will neither be bullied and certainly not invaded.

The Middle East remains a tinder box with multiple uncertainties, as do the various global terrorist threats, the rivalries between nations such as Pakistan, India, and China, and if, how and when more belligerent, neo-fascist, extreme right-wing movements develop into a range of potential threats to world stability.

Sanctions have limited deterrence effects, as has been seen during the Ukraine War, with negative impacts on those who have imposed well-intentioned sanctions. At the end of the day there is no substitute for a hugely strong, highly capable alliance-based deterrence strategy that signals that outright aggression will not work.

The USAF's strategic significance in the context of its enduring presence in the UK for the foreseeable future now becomes clearer in terms of the options open to the UK to decide and implement priorities. The USAF capabilities are, very simply, prodigious. Most important of all they compose a critical component of the global deterrence strategy described earlier. What is now required is a clear and concise vision of how the UK should adjust to these capabilities and tailor its defence programmes and operations in ways that will optimise its contributions to a deterrence strategy in conjunction with its principal ally, the United States, and the other key allies within the Five Eyes Intelligence community, NATO, and further afield, such as Japan and India.

The strategic significance is this. The United Kingdom now has a critical and timely opportunity to address how to best contribute and then how to allocate resources and manpower, including programmes that will fit best into the global allied deterrence tapestry.

# Chapter 20

# A Way Ahead for the UK and US in Air Warfare and its Impact on UK Policy, Operations, Acquisitions and Defence Structure

The Falklands War demonstrated at strategic and tactical levels the multiple virtues and lasting value of naval power. Sharkey Ward's detailed descriptions and analysis demonstrated beyond all reasonable doubt that the United Kingdom requires a larger and more capable navy with naval air capabilities that will enable the UK to fully be interoperable and interchangeable with its principal ally, the United States, and specifically the United States Navy and the United States Marine Corps.

Sharkey's chapters pave the way for the present and future development of a much more capable Royal Navy aviation component that is formidable and can operate truly as an equal air arm with the United States Navy.

How can this be achieved in a UK that is undergoing a variety of budgetary stresses and strains, and where defence spending has to be carefully measured?

The strategic context has been described earlier. The UK can maximise its contribution to an alliance-based global deterrence posture by not only endorsing but also actively pursuing a maritime strategy.

The way to achieve this and indeed the only true way ahead is provided by no other than the United States Air Force.

The foregoing chapters have shown how the continuity of USAF presence and capabilities in the UK present the UK government with options to concentrate on a maritime strategy and, in doing so, expand naval air capabilities.

The air defence of the UK from Russian intrusions is a task that can be met by a limited, scaled-down Royal Air Force order of battle. In addition, Royal Navy aircraft can support this particular mission. The presence of an increasingly capable USAF in the UK provides the UK government with a significant opportunity, a true moment in time, to make a critical change in its defence budget and programmes. The Royal Air Force can never match the numbers and capabilities of the United States Air Force. However, it can contribute in many ways that are invaluable, and these were described earlier. That is the way ahead for the Royal Air Force.

In the financial year 2022 the United States Air Force annual budget was $213 billion. The entire defence budgets of the United Kingdom, Germany, France, and Japan in FY 2022 was $213.8 billion. These numbers are a serious reality check for the United Kingdom. Chris Cope, a distinguished UK defence analyst and correspondent, pointed out to the authors that the proposed Tempest programme for the Royal Air Force of approximately £17 billion over the life of the programme, at today's costs, could potentially cripple the UK defence budget. Cope pointed out that although this programme is a proposed UK-Italy-Japan endeavour the potential competition with the US F-35 programme may well cause disarray with the United States. Furthermore, there is a possibility that the White House and its key defence advisors in the Department of Defense and the National Security Council may conclude that the United Kingdom is backtracking on its commitment to the United States Indo-Pacific Command maritime deterrent strategy based on alliance cohesiveness.

The resulting reduction in the Royal Air Force budget, and totally forgoing any plans for a new expensive RAF fighter replacement programme, Tempest, enables the UK to direct resources to a maritime strategy in close collaboration with the United States.

The budget changes would enable the UK to make several key acquisitions that will enhance its roles and missions in the global commons.

The Royal Navy would be able to fully populate its two aircraft carriers with a full complement of F-35Bs. These aircraft are expensive, and it behoves the UK to show total commitment and resolve to a maritime strategy by the acquisition of these critically needed aircraft.

Over the long term, as the UK defence budget is restructured, together with the necessary manpower changes in the Royal Air Force and Royal Navy, the UK may plan and execute an affordable naval expansion. This can only be achieved in the current and projected UK defence budget by a comparable reduction in the budget of the Royal Air Force. The latter's aircraft inventory is not just ageing and costly in terms of maintenance, it is also increasingly less viable in the current and projected air warfare threat environments. The UK-based USAF assets are and will be much more capable and relevant. The UK would acknowledge reality by concentrating on one critical domain that also meets its own vital national security interests in addition to those of its principal allies. The UK would in effect return to its historic roots in an ever-changing global threat environment, namely a maritime strategy that underpins and protects the vital national security interests of its people and its global interests. Reality will have been faced and, in so doing, institutional loyalties and service rivalries will have been put aside in the national interest.

Over the long haul the UK would be able to consider other investments to augment its maritime strategy in conjunction with the United States, once the above is stabilised and budget issues resolved.

There will be many opportunities. For just one example, the UK would in due course be able to consider and afford reconfiguring the flight deck of its current aircraft carriers and/or building another aircraft carrier that is capable of launching and retrieving both conventional fixed-wing as well as short take-off aircraft such as the current F-35B. This would require an angled deck with catapult launch and arrester systems. The latest US Navy nuclear-powered aircraft class, *Gerald Ford*, has an Electromagnetic Aircraft Launch System (EMALS), replacing the steam catapult system that has been in earlier British aircraft carriers and the current US Navy *Nimitz*-class nuclear powered aircraft carriers (CVNs). The Royal Navy could then operate both aircraft types and realise the status of having available true strike carrier capability.

The future roles and missions of the Royal Air Force were described earlier. Though the traditional missions will disappear, others remain, and new critical ones emerge. The air refuelling and transport missions remain, and the intelligence and specifically space-related operations in conjunction with the newly established United States Space Command can be augmented to meet the dramatically changing threat environment. For example, space is a vital dimension in addressing the threat from hypervelocity missiles. Detection, tracking and engagement of such threats will depend in part on space-based systems and technologies.

The United Kingdom has much to contribute to an alliance-based, deterrence-oriented strategy that is underscored by war-fighting capabilities that continuously signal potential threats that aggression will not work or be sustainable in the event of conflict.

There are therefore many achievable opportunities for the United Kingdom to focus its defence resources on ways that will maintain its standing as a key dependable and capable ally and contribute significantly to the defence of its vital national interests in conjunction with those of its principal allies.

It is time for visionary change, not falling back on well-worn procurement paths and possibly responding to a mixture of industrial base interests and service rivalries and loyalties that may be counterproductive. Integrity and innovation should characterise the way ahead, not dependence on an out-of-date and outmoded defence culture and system. Evolving naval power will endure in parallel with, for example, the evolution of space-based intelligence, surveillance and reconnaissance.

It is a time for positive change. The United Kingdom has the inherent ability to grasp the opportunities presented and to move forward, onwards and upwards.

# Epilogue

In warfare the hope is always that the past will not be prologue. Those who gave their all would have wished that their sacrifices were not in vain, and that the world would learn and be a better place. Sadly, the world has often not learned from past lessons, and particularly from the horrors, trials, and tribulations of the Second World War. The democracies have struggled valiantly since 1945 to make the world a better place through institutions such as the United Nations, and all its many agencies, the North Atlantic Treaty Organisation and, at another level, organisations such as the Five Eyes Intelligence community, which is committed to peace and security through sound, reliable, and timely intelligence to warn and indicate security challenges.

Deterrence has been emphasized as the best precaution against impending or actual threats. In order to deter war, the US and the UK must have superior military capability, demonstrating strength through power, and also the overarching capacity to endure a conflict of whatever dimensions and for whatever period. Risk is an ever-present factor. The US and the UK have to stand their ground together against those threats that threaten to destabilise the world and challenge the democratic way of life. Capabilities are critical to face down an emergent China and a dictatorial Russia. Fighting a war of attrition against a potential enemy such as China requires serious capabilities.

Allies are critical. A country as strong and as powerful as the United States cannot fight alone. The United Kingdom is its principal ally. The US-UK military and intelligence special relationship is unambiguous. It is solid and enduring. The UK now needs to be flexible and willing to commit to change that is in part cultural as well as operational, with investment and acquisition issues strikingly relevant. The UK simply cannot afford to be an ally that does not invest its precious defence pounds wisely.

Flexible thinking, innovation, and commitment are now required to address what may be called 'the conclusion of the whole matter' regarding the future defence posture of the United Kingdom apropos its critical alliance with the United States. Organisational change should be driven by the arguments put forward for a predominantly UK maritime strategy in which naval air power and submarine forces play key roles alongside surface forces and the enduring capabilities of the Royal Marines.

A new balance has to be established, one that should witness a reduction in the tactical aircraft elements of the Royal Air Force in favour of naval air power, including the Poseidon P-8 maritime anti-submarine and reconnaissance aircraft that should be transferred to Royal Navy command and control, and maintenance. Those domains where the Royal Air Force may contribute optimally have been described in the preceding chapters. Space, intelligence, cyber, tankers, transport, and in-flight refuelling are critical warfare functions that the Royal Air Force can perform with great effect and distinction.

But global readiness and agility are critical. Potential adversaries have to be kept guessing. The Royal Navy is a forward-deployed force. It can deploy globally. Together with the fleets and units of the United States Navy, the Royal Navy and the Royal Marines must have the superior capability and sufficient capacity to endure a lengthy conflict. The naval air capacity of the United Kingdom has to be increased considerably as an urgent priority. The credible capacity to fight and be a reliable ally is dependent on the UK government making major budgetary changes in accordance with the cultural and organisational changes required of a national maritime strategy.

The United Kingdom is at an inflection point. If it does not accelerate change it may possibly become an ineffective ally. Transformation is achievable with good leadership and political will. The Indo-Pacific challenge alone should be a wake-up call to the UK's leadership.

More investment in the war-fighting strength, flexibility, survivability, mobility, and adaptation of UK naval air power would add a huge capability to a totally collaborative alliance with the United States.

The history of the United States Air Force in the United Kingdom will enter a new era. This will reflect the enduring traditions and capabilities that have been described in the early chapters. Whether it is the presence of the new B-21 bomber, or enhanced space-based tracking with, for example, more resilient missile warning capabilities in the UK, while continuing to be a forward-deployed major deterrent force in NATO based in the UK, the USAF presence permits the United Kingdom leadership to make the critical changes described above.

There is a clear and unequivocal need for change. Let the United Kingdom reflect on the fine past achievements and sacrifices described in these chapters as part of a continuum of positive change that will take the United Kingdom with its great ally, the United States, into a new era.

# Authors' Biographies

## Commander N.D. MacCartan-Ward DSC AFC, Royal Navy; better known as 'Sharkey' Ward

Sharkey on deck after a Falklands mission. Painting by Leanne Gilroy.

## Background – the Early Years

Sharkey was brought into the world as Nigel David Ward on 22 September 1943 in Medicine Hat, Alberta, Canada, where his RAF father was serving. (He changed his surname by deed poll in the early 1990s to MacCartan-Ward after his father's death and in honour of the latter's Irish ancestry.) He travelled to the UK in 1944 with his mother and elder brother – a five-week journey by sea, avoiding the U-boat threat. He was stricken with bronchiectasis and hospitalised for a year at the age of 5, and only survived the disease by spending three years in the dry climate of Pakistan at RAF Mauripur, Karachi. After a return to England, he was educated as a boarder at Reading School, Berkshire (1954–62), where he became head boy and captain of the 1st XV rugby team (and also a South of England rugby trialist.) He learned to fly with an RAF flying scholarship in 1959, earning his private pilot's licence in the Tiger Moth.

## Royal Navy Career – as a Seaman Career Officer

In 1962, he entered Britannia Royal Naval College as a naval cadet on the general list.

In 1963–64 he served as a midshipman in the Bahamas Patrol frigate HMS *Tartar* before returning to the college as a sub-lieutenant. He was put in charge of the Dartmouth Tiger Moth Flying Club and appointed divisional sub-lieutenant of Exmouth Division. On leaving Dartmouth, the Commander of the College, John Ford, commented that: 'Never has a Division been run so well from a horizontal position!'

In 1965–66, as a lieutenant, he was appointed Navigator of the Hong Kong Guardships, HMS *Penston* and *Woolaston*. During this period, he achieved his bridge watch keeping and ocean navigation certificates and qualified as a ship's diver.

## Royal Navy Career – Flying Training

Sharkey's military flying training began in 1966 and continued until 1969. When passing out from basic flying training, flying the Jet Provost at RAF Linton-on-Ouse, he won the aerobatics trophy and the Lord Mayor's Sash for best student. His advanced and tactical flying training took place at Royal Naval Air Station Brawdy in South Wales where he flew the Hunter fast jet aircraft. His follow-on operational flying training took place at Royal Naval Air Station Yeovilton in Somerset, where he flew the front-line fighter aircraft the Sea Vixen Mk.2 and where he was awarded the Kelly Memorial Trophy and Admiral Sir Dudley Pound Prize for best operational student. He was then appointed to the Phantom training squadron, 767 NAS, and to 892 NAS, a front-line F-4K Phantom unit, in continuation.

## Royal Navy Career – in the Front Line with the Fleet Air Arm

Sharkey carried out his first deck landing in HMS *Ark Royal* on 14 June 1970. He qualified as an all-weather fighter pilot by day and night. He then qualified as an instrument rating instructor in 1971 and an air warfare instructor in 1972.

From 1972 to 1974, he carried out a NATO aviation staff job as Nuclear Intelligence and Planning Officer at Allied Forces Northern Europe (AFNORTH), Oslo, Norway, under the command of General Sir Peter Whitely, Royal Marines. While there, he drafted: the mine warfare policy for AFNORTH; the plans for the nuclear destruction of the Soviet amphibious fleet; and helped draft a new 'graduated response' nuclear policy for the Supreme Allied Commander Europe (SACEUR) to replace the existing 'tripwire' policy, which was accepted.

On return to the UK, he was appointed as the senior pilot of 892 Squadron and strove successfully to enhance the night deck landing and air-to-air combat confidence and capability of all his fourteen pilots (including himself).

His second embarked front-line tour ended in 1976 when he was appointed to serve in the Ministry of Defence as the Desk Officer in charge of the Sea Harrier FRS Mk.1, the P3T Sea Eagle sea-skimming missile and the AIM-120B AMRAAM projects. By 1979, the Sea Harrier was ready to enter service, on cost and on time. He was then appointed to be the Commanding Officer of the Sea Harrier Intensive

Flying Trials Unit, 700A Squadron, was ordered to write his own terms of reference and, with an excellent team of aviators and engineers, was relied upon entirely to prepare the aircraft and all its personnel for full operational readiness. In 1980, he was appointed to command the Headquarters Squadron, 899, in continuation and, in 1981, was given command of 801 Squadron based in HMS *Invincible* and was promoted to commander (by which time he had become well known as 'Mister Sea Harrier').

During Operation Corporate, the Falklands War, he was senior Sea Harrier advisor to the Command on all aspects of the fast jet air war, flew over sixty war missions by day and night, achieved three air-to-air kills (and one damaged) and was involved in/witnessed eight further kills.

HMS *Invincible* being put through her paces.

During the Falklands War, he was awarded the Air Force Cross for Services to Harrier Aviation and then, after the war, he was decorated with the Distinguished Service Cross for Gallantry.

Having returned to the UK, he received the Freedom of the City of London, became a Member of the Royal Aeronautical Society and achieved first place at Greenwich Staff College, receiving the director's prize. He then served in the Ministry of Defence as the Air Warfare and Air Weapons Adviser to the Naval Staff and the First Sea Lord before voluntarily retiring from the Royal Navy in 1985.

## Falklands-related Accolades

During 1976, one of his Phantom pilots: 'Sharkey, you might be a crazy bastard, but if we ever go to war, I want it to be with you!'

## Post-war

First Sea Lord, Admiral Sir Henry Leach: 'Without the Sea Harrier there could have been no Falklands War.'

Admiral Sir 'Sandy' Woodward GBE KCB, Commander of the Carrier Battle Group, Falklands 1982: 'If Sharkey Ward had not disobeyed orders, we would/could not have won the Falklands war.'

Admiral Sir Jeremy Black GBE KCB DSO, Captain of HMS *Invincible* in the Falklands:

While everyone on board played a significant role, not least 820 squadron with nine Sea Kings – they flew the equivalent of once around the world often in appalling weather without the loss of an aircraft – it must be 801 Squadron, the Sea Harriers, who take the ship's honours for the campaign. This small body of men, under Lieutenant Commander 'Sharkey' Ward, faced the enemy at close range. They had been reinforced by some aircraft of 899 Squadron. They shot down nine aircraft, of which Sharkey shot down three, including a Hercules transport.

Admiral Sir Derek Roy Reffel KCB, Flag Officer Flotilla 3, who signalled: 'No other ship has contributed so much to the success of the Task Force group operations [as *Invincible*].'

Major-General Julian Thompson CB OBE, Brigadier commanding 3 Commando Brigade in the Falklands: 'Sharkey Ward's book [*Sea Harrier over the Falklands*] will come as a revelation to anyone who thinks that a fighter pilot's prelude to battle is a matter of kicking the tyres, lighting the fires and heading for the wide blue yonder. Painstaking preparation of his squadron, superlative personal flying skills, and a determination to engage the enemy more closely, made Sharkey Ward a fighter leader in the mould of Bader, Johnson, Ball and others in the past.

To those of us who day after day watched the air attacks on the ships in San Carlos, the best news was the number of Argentine aircraft splashed after each raid. As the Sea Harriers whittled down the enemy, so our admiration for the Fleet Air Arm increased. Without them we would not have won.'

Prime Minister Margaret Thatcher: 'Sharkey Ward is one of the six most important names of the Falklands War.'

A private message to Sharkey from one of his squadron maintenance engineers: 'Sharkey, we would gladly have died for you.'

## Beyond the Falklands

### 1985

Managing director of Defence Analysts Limited:

- Consultant to British Aerospace and Ferranti Radar.
- Provided security services to an international range of clients including Petroleum Pipeline of Panama and the Thai government.
- Provided on board safety, defence and emergency operators for oil tankers and special cargo vessels in the Arabian Gulf during the Iran-Iraq War. Saved many ships and many lives.
- Working with UK government politicians and scientific advisers to set up a fishery protection service for the Sierra Leone government. Drew up all the plans and procedures and raised financial support from Citibank to the tune of £17 million. Political troubles within

Sierra Leone and neighbouring Liberia eventually prevented the project from proceeding.

**1998 to 2018**

Since the turn of the century, Sharkey has worked hand in glove with Admiral Sir Sandy Woodward (RIP) to address the vital needs of the UK's strategic maritime policy – with particular emphasis on providing naval air warfare and aircraft carrier operations expertise to the Ministry of Defence and associated government committees.

## Books

*Sea Harrier over the Falklands*. Published by Pen and Sword and presented by them as a Military Classic. E-book and audiobook available (www.hermajestystopgun.com).

Review by publishing editor, Toby Buchan, grandson of John Buchan, author of *The 39 Steps*: 'I have been editing military books for more than 20 years and this is the best book of its kind that I have ever read.'

*Her Majesty's Top Gun and the Decline of the Royal Navy*. Published by Mill City Press. E-book and audiobook available.

Reviewed by Dr Anthony Wells: 'Sharkey Ward is a fine Gus Halliday look alike. They were made in the same mould. They were both distinguished naval aviators, both saw what was needed and why, and both took different steps to make things worked at their different levels and roles of responsibility.

'His long-awaited new book, "Her Majesty's Top Gun", is out, coming at the right time to set the record straight. It is a great read. I was privileged for Sharkey to invite me to read his terrific manuscript. It's a wonderful epic story. In my humble opinion Sharkey is the UK's most distinguished naval aviator since World War II.

'Every member of Britain's Parliament, government, and all the Five Eyes allies should read "Her Majesty's Top Gun", together with the allies' defence and security communities.

The layperson will enjoy this book too. It is gripping.

'Gus Halliday would applaud.'

# Dr Anthony R. Wells

Dr Anthony R. Wells (seen in Prague).

Anthony Wells is unique insofar as he is the only living person to have worked for British intelligence as a British citizen and US intelligence as a US citizen, and to have also served in uniform at sea and ashore with both the Royal Navy and the US Navy. He is a fifty-year veteran of the Five Eyes intelligence community. In 2017 he was the keynote speaker on board HMS *Victory* in Portsmouth to commemorate the 100th anniversary of the famous Zimmermann telegram intelligence coup by 'Blinker Hall' and his Room 40 team in British Naval Intelligence. The guest of honour was Her Royal Highness Princess Anne, with the Five Eyes community, past and present, represented from the United States, the United Kingdom, Canada, Australia, and New Zealand.

Dr Wells, or Commander Wells, was trained and mentored in the late 1960s by the very best of the Second World War intelligence community, including Sir Harry Hinsley, the famous Bletchley Park code breaker, official historian of British Intelligence in the Second World War, Master of St John's College, Cambridge, and Vice Chancellor of Cambridge University. Sir Harry Hinsley introduced Dr Wells to the Enigma data before it became public knowledge. Dr Wells received his PhD in War Studies from King's College, University of London, in 1972. He holds bachelor's and master's degrees from the University of Durham, and a master's degree from the London School of Economics. He was trained at Britannia Royal Naval College, Dartmouth, and received his advanced training at the School of Maritime Operations. He was called to the Bar by Lincoln's Inn in November 1980.

The 2017 HMS *Victory* Intelligence Community Dinner.

Anthony Wells has four children and eight grandchildren, and lives on his farm in Virginia. He is a Member of the Naval Order of the United States and was appointed an Honorary Crew Member of USS *Liberty* by the USS Liberty Veterans Association. USS *Liberty* is the most highly decorated warship in the history of the US Navy for a single action, attacked by Israeli air and surface forces on 8 August 1967, in the eastern Mediterranean. Dr Wells is the third Chairman of the USS Liberty Alliance, succeeding the late Admiral Thomas Moorer, former Chairman of the US Joint Chiefs of Staff and Chief of Naval Operations, and the late Rear Admiral Clarence 'Mark' Hill, former distinguished US naval aviator and battle group commander. He is a retired US national ski patroller and instructor, and a life member and former President of The Plains, Virginia, Volunteer Fire and Rescue Company. Wells is an FAA commercial pilot with single and multi-engine, land and sea, instrument, and flight instructor ratings.

Dr Wells was the Technical Director of Fleet Battle Experiments ALPHA and BRAVO in the Third Fleet, United States Pacific Fleet. He was the Chief Executive Officer of TKC International LLC, a specialist company supporting the US Intelligence community and Department of Defense, for twenty-five years. He held top secret SCI and special access clearances.

## Anthony Wells' Publications

### Literary Awards

In 2013 and 2017 the United States Submarine League presented Dr Anthony R. Wells with Literary Awards for Articles in *The Submarine Review*.

### Books

*German Public Opinion and Hitler's Policies, 1933–39*. 1968. Electronic version available at Durham University Library, UK – access Durham University Library online and enter database with title and/or author name. Electronic and hard copy versions available.

*Studies in British Naval Intelligence, 1880–1945*. 1972. Electronic version available online via the British Library (ETHOS), and also online at King's College, London by entering the database with title and/or author name. Electronic and hard copy versions available.

*Training and the Achievement of Management Objectives, the Solution of Management Problems, and as an Instrument of Organisational Change*. 1974. The London School of Economics and Political Science.

*Technical Change and British Naval Policy*. Edited by Bryan Ranft, Hodder and Stoughton, London, 1977, and Holmes and Meier, New York, NY.

*War and Society*. Edited by Brian Bond and Ian Roy, Croom Helm, London, 1977, and Holmes and Meier, New York, NY.

*Soviet Naval Diplomacy*. Edited by B. Dismukes and J. McConnell, Pergamon Press, 1979.

*The Soviet and Other Communist Navies*. Edited by James George, US Naval Institute Press, Annapolis, Maryland, 1986.

*Black Gold Finale.* A novel. Dorrance Publishing Company, 2009.

*The Golden Few*. A novel. Dorrance Publishing Company, 2012.

*A Tale of Two Navies: Geopolitics, Technology, and Strategy in the United States and the Royal Navy, 1960–2015*. US Naval Institute Press, Annapolis, Maryland, January 2017.

*Between Five Eyes*. Casemate Publishers, Oxford, UK & Havertown, Pennsylvania, September 2020.

*Room39 and the Lisbon Connection.* A novel. Xlibris, Bloomington, Indiana, June 2021.

*Crossroads in Time Philby & Angleton. A Story of Treachery*. A novel. Palmetto Publishing, Charleston, South Carolina, 2022.

*Gone to Earth A Young American Woman Disappears in the South Pacific. Based on a True Story*. A novel. Xlibris, Bloomington, Indiana, 2022.

**Articles**

Admirals Hall and Godfrey – Doyens of Naval Intelligence (Two Parts). *The Naval Review*, 1973.

Staff Training and the Royal Navy (Two Parts). *The Naval Review*, 1975, 1976.

The 1967 June War: Soviet Naval Diplomacy and the Sixth Fleet – A Reappraisal. Centre for Naval Analyses, Arlington, Virginia. Professional Paper 204, October 1977.

The Center for Naval Analyses. Professional Paper Number 197, December 1977. Department of the Navy, Washington DC, Center for Naval Analyses.

The Soviet Navy in the Arctic and North Atlantic. *National Defense*, February 1986.

Soviet Submarine Prospects 1985–2000. *Submarine Review*, January 1986.

A New Defense Strategy for Britain. *Proceedings of the United States Naval Institute*, March 1987.

Presence and Military Strategies of the USSR in the Arctic. Quebec Center for International Relations, Laval University, 1986.

Real Time Targeting: Myth or Reality. *Proceedings of the United States Naval Institute*, August 2001.

Missing Magics Machine Material. New Insights on 7 December 1941, and Relevance for Today's Navy. *The Submarine Review*, April 2003.

US Naval Power and the Pursuit of Peace in an Era of International Terrorism and Weapons of Mass Destruction. *The Submarine Review*, October 2002.

Transformation – Some Insights and Observations for the Royal Navy from Across the Atlantic. *The Naval Review*, August 2003.

They Did Not Die In Vain. USS Liberty Incident – Some Additional Perspectives. *Proceedings of the United States Naval Institute*, March 2005.

Royal Navy at the Crossroads: Turn the Strategic Tide. A Way to Implement a Lasting Vision. *The Naval Review*, November 2010.

The Royal Navy is Key to Britain's Security Strategy. *Proceedings of the United States Naval Institute*, December 2010.

The Survivability of the Royal Navy and a New Enlightened British Defense Strategy. *The Submarine Review*, January 2011.

A Strategy in East Asia that can Endure. *Proceedings of the United States Naval Institute*, May 2011.

A Strategy in East Asia that can Endure. *The Naval Review*, August 2011. Reprinted by kind permission of the United States Naval Institute.

The United States Navy, Jordan, and a Long-Term Israeli-Palestinian Security Agreement. *The Submarine Review*, spring 2012.

Admiral Sir Herbert Richmond: What would he think, write and action today? The Naval Review, February 2013 – Lead article in the Centenary Edition of *The Naval Review*.

Postscript to Missing Magics Machine Material – Tribute to a Great Submariner: Captain Edward Beach, US Navy. *The Submarine Review*, 2013.

Jordan, Israel, and US Need to Cooperate for Missile Defense. USNI News, 26 March 2013.

A Tribute to Admiral Sir John 'Sandy' Woodward. *USNI News*, 8 August 2013.

USS LIBERTY Document Center. Edited by Anthony Wells and Thomas Schaaf. A website produced by SiteWhirks, Inc., Warrenton, Virginia, September 2013. In April 2017 this website was transferred to the Library of Congress for permanent safekeeping for the use of future scholars and researchers.

The Future of ISIS: A Joint US–Russian Assessment. With Dr Andrey Chuprygin. *The Naval Review*, May 2015.

The Zimmermann Telegram: 100th Anniversary. *The Naval Review*, February 2017 and *The Submarine Review*, 2017.

Put The Guns in a Box: With Captain J.W. Phillips, US Navy retired. *Proceedings of the US Naval Institute*, June 2018.

Quo Vadis China? A View from Across the Atlantic. Part 1. *The Naval Review*, November 2019.

Quo Vadis China? *The Submarine Review*, December 2019.

USS Amberjack and the Attack on USS Liberty: With Mr. Larry Taylor, ST1 USS Amberjack. US Naval Institute Naval History Blog, 7 January 2020.

USS Amberjack & the Attack on USS Liberty. With Mr. Larry Taylor. *The Submarine Review*, March 2020.

The UK's Strategic Defence & Security Review, A US Perspective. *The Submarine Review*, June 2020.

The United Kingdom Needs a Maritime Strategy. *The Naval Review*, August 2020.

Submarines and the Ring of Fire in the Indo Pacific Theater: A Strategic Analysis. *The Submarine Review*, December 2020.

UK's Defence & Security Review – Some Final Observations: *The Naval Review*, Autumn 2020.

A Brave New World of Next Generation Technologies: *Warship World*: Volume 17, Number 2, January/February 2021.

To Honor The Last Nuremburg Prosecutor: *Proceedings of the United States Naval Institute*, May 2021, Annapolis, Maryland.

The United Nations Convention on the Law of the Sea and the United States Navy: US Naval Institute Blog, June 2021.

Is There a Need for a New Generation of Submarine Officers who are Intelligence Trained and Experienced beyond Current Levels? & How Might We Learn from the Past? *The Submarine Review*, June 2021.

Behind the Five Eyes: *Counsel Magazine* (Justice Matters: Spotlight section), the monthly magazine of the Bar of England & Wales, London, UK, July 2021.

Letter from The Plains: A monthly article in the Middleburg Eccentric, Virginia, since 2016.

**Reports**

NATO and US Carrier Deployment Policies. Center for Naval Analyses, Arlington, Virginia, February 1977.

NATO and US Carrier Deployment Policies, Formation of a New Standing Naval Strike Force in NATO. Center for Naval Analyses, Arlington, Virginia, April 1977.

Sea War '85 Scenario. With Captain John L. Underwood, USN. Center for Naval Analyses, Arlington, Virginia, June 1977.

Submarine Construction programme for the State of Sabah, Malaysia. RDA Contract TR-188600-OOl, December 1984. Chief Minister of Sabah, Malaysia and Government of Malaysia.

The Application of Drag Reduction and Boundary Layer Control Technologies in an Experimental programme, January 1985. For the Chief Naval Architect, Vickers Shipbuilding and Engineering Ltd, Barrow-in-Furness, UK.

The Strategic Importance and Advantages of Labuan, Federal Malaysian Territory, as a Naval Base with Special Reference to its Capabilities as the Royal Malaysian Navy Submarine Base, March 1985. Chief Minister of Sabah, Malaysia and Government of Malaysia.

Preliminary Overview of Soviet Merchant Ships in Anti-SSBN Operations and Soviet Merchant Ships and Submarine Masking. (Department of the Navy Contract N00016-85-C-0204).

SSBN Port Egress and the Non-Commercial Activities of the Soviet Merchant Fleet: Concepts of Operation and War Orders for Current and Future Anti-SSBN Operations. (Department of the Navy Contract 136400).

Overview Study of the Maritime Aspects of the Nuclear Balance in the European Theater (Department of Energy Study for the European Conflict Analysis Project), October 1986.

Soviet Submarine Warfare Strategy Assessment and Future US Submarine and Anti-Submarine Warfare Technologies (Defence Advanced Research Projects Agency, March 1988), RDA Contract 146601).

Limited Objective Experiment ZERO, July 2000. The Naval Air Systems Command, US Navy, Department of Defense, 2002.

Operational factors Associated with the Software Nuclear Safety Analysis for the UGM-109A Tomahawk Submarine-Launched Land Attack Cruise Missile Combat Control System Mark I. United States Navy and Logicon Inc., 1989.

Operation Bahrain, March 2003. The Assistant Director of Central Intelligence, the Central Intelligence Agency.

Distributed Data Analysis with Bayesian Networks: A Preliminary Study for Non-Proliferation of Radioactive Devices, December 2003 (with F. Dowla and G. Larson). The Lawrence Livermore National Laboratory, Livermore, California, December 2003.

Fiber Reinforced Pumice Protective Barriers – To Mitigate the Effects of Suicide and Truck Bombs. Final Report and Recommendations. United States Navy, Washington DC. With Professor Vistasp Kharbari, Professor of Structural Engineering, University of California, San Diego, August 2006.

Weapon Target Centric Model: Preliminary Modules and Applications, in Two Volumes. United States Navy, Principal Executive Officer Submarines, Washington DC, August 2007.

Tactical Decision Aid (Tda): Multi intelligence capability for National, Theater, and Tactical intelligence in real time across geographic space and time. The National Intelligence Community, Washington DC, May 2012.

Submarine Industrial Base Model: Key industrial base model for the US Virginia Class nuclear-powered attack submarine, Principal Executive Officer Submarines, Washington Navy Yard, Washington DC, October 2012.

**Manuals**

Astro-Navigation: A Programmed Course in Six Volumes for Training UK and Commonwealth Naval Officers in the Use of Astronomical Navigation at Sea. Royal Navy, Ministry of Defence, UK, 1969.

The Battle of Trafalgar: A Programmed Course in one Volume in Naval Strategy and Tactics. Royal Navy, Ministry of Defence, 1969.

The Double Cross System: A Programmed Course In one Volume for British, Foreign and Commonwealth Naval Officers Attending the Royal Naval Staff College, Greenwich, UK. Royal Navy, Ministry of Defence, 1973.

**Unclassified Titles for Technology and Operational Areas – Covering Classified Programmes – and Publications – Generic Areas**

Airborne Mine Clearance

Streak Tube Imaging LIDAR

Magic Lantern programme

Tritium Micro sphere Technology

Classified Applications of the Naval Simulation System

Naval Surface Fire Support and the Extended Range Guided Munition (ERGM)

Non-Acoustic Antisubmarine Warfare

Battlefield Awareness and Data Dissemination (BADD Programme)

Joint Stars Programme Special Applications

Naval Fires Network

Littoral Surveillance System

Fleet Battle Experiment Operations (Technical Director FBE Alpha and FBE Bravo) Third Fleet, US Pacific Fleet

Ocean Surveillance (radar and optics)

Multi Spectral Applications

Space Based Sensors and Surveillance

Microwave Radiometry Applications

Detection, Locating and Tracking

Clandestine Operations and Intelligence Collection Operations

Support to Special Forces

Special Submarine Operations

Tagging Tracking and Surveillance

Battlespace Shaping and Real Time Targeting

Covert and Clandestine Operations against Weapons of Mass Destruction and
Other major threats to US Security
Special Sensor Technology
Covert & Overt Operations Planning and Execution
Reports and MOUs for Commander-in-Chief and Secretary level actions
Airborne Infrared Measurement System
Stealth and Counter Stealth
Counter-Intelligence Operations
Tactical Exploitation System and Joint Fires Network
Asymmetric Warfare Initiative – 2003
Hairy Buffalo Programme
Tracking of the al Qai'da Terrorist Network and Operations
Tactical Decision Aid (TDA) for Submarine ISR operations
Advanced Cyber Attack and Defense Technologies and Operations
Shrouded Lightning Special Programme
Non-Linear Junction Radar and Adaptive Regenerative Controller Special Programme
Special Programme in Jordan
Special Programme in Malaysia
Special Programme in Bahrain
Special Programme in Abu Dhabi
Special Programme in Saudi Arabia
Special Programme with Commander United States Pacific Fleet
Special Tests at the US Naval Air Station Patuxent River, Maryland, September 2012
LISAC Special Programme
Applications of the Robust Laser Interferometry (RLI) system and technology
Special Support to a combined Cheltenham UK and Maryland US Group
Special Support for Indo-Pacific Operations

**Classified Titles and Publications**

1968–2018 Dr Wells has been the author, lead author, or a key author of multiple highly classified Codeword documents at the Top Secret SCI level in both the United Kingdom and the United States.

# Index